HOW TO NAME YOUR CARBON COMPOUND

VOL 1

PURANJAY SRIRAMBHATLA

2nd year Integrated M.Sc (2022-23)
NIT Warangal

Contents

Preface

This is a layman's book to understand a part of Organic chemistry. This book consists of the basics which will help in understanding other depths. It covers How to name different organic compounds, functional groups and other aspects of the subject called "organic chemistry". The language used in this book is simple and easy to understand and contains elaborately explained theories and other concepts. The purpose of the book is to encourage and enhance the understanding of the beauty and mysteries of organic chemistry. Chemistry is subject which always finds a better way to answer a problem; as a student of science, we will have to find our own path to answer a question. For that, we will also have to know the answers given by others; that is when your answer will seem perfect until someone else answers better than you.

<u>Thanks</u>

To: All the science teachers who have guided me and inspired me to take up such a job and put in the place where I am today.

<u>Special Thanks</u>

To: _<u>Dr. V.Rajesh Kumar</u>_, Assistant Professor, Department of Chemistry, NIT Warangal.

Thank you sir for helping me to complete this book and guiding me throughout and correcting me when I stubble with some error.

To: _<u>Ajeesh A K</u>_, Reasearch scholar NIT Warangal

Thank you sir for taking time and verifying my text.

The existence of the person depends on the way he lives and the work he does, the existence of a tool depends on its usage in running the clock of time, and the existence of a creature depends on the survival of the fittest.

All the things mentioned above perish with time. That is, a human will die at a point in time, the tool becomes useless when there is an alternate replacement for it, and a creature becomes prey if it does not adapt to its surroundings in the process of survival.

As **Albert Einstein** said, energy could neither be created nor destroyed. In the same way, when a human perishes, his body will become a host for many other organisms and help in their growth and mutation. In the same way, metal can be recast into parts of advanced machinery, but it does not change its properties.

When we talk about existence, one should never forget to mention carbon, an essential element of our very existence. It is present in every cell of the human body with different functions and purposes. It is also present in the largest tree to the smallest herb. It is in the ashes of a volcano to the most expensive diamonds. From the deadly gas like carbon monoxide to the daily used fossil fuel. Everything contains carbon.

How can a single atom have these many forms?

How much is there about carbon we don't know?

How do we know about carbon present in our own bodies?

Is this the base of the vast subject called organic chemistry?

These are just the beginning of the question on carbon. How many questions are known, how many of them are answered and how many are left unanswered?

Chapter 1

Introduction to Carbon

Let's start from the beginning. Although charcoal, soot and diamond were known to human civilization for a long time, it was until 1722 that **Antoine Ferchault** demonstrated the existence of an element now called carbon which is important to convert iron into steel.

Antoine Ferchault proved that charcoal and diamond are different forms of carbon by oxidizing them with oxygen to produce carbon dioxide.

Later in 1786, a group of French scientists confirmed that graphite is also a form of carbon by oxidizing it in the same way as Antoine Ferchault. Carbon as an element was first printed in the 1789 textbook by **Antoine Lavoisier**; this is how the journey of carbon and its compounds began.

Later in 1865, John Alexander Newlands introduced carbon in the periodic table, known to us as the laws of octaves. He observed that when elements were arranged in increasing atomic mass order, every eighth element had similar property as the first.

But he was unsuccessful because he could not arrange all elements discovered during his period.

Later in 1869, Dmitri Mendeleev proposed a newer method of distinguishing elements based on their atomic masses and their reactivity. He has classified the elements on the bases of their reaction with hydrogen and oxygen.

He has observed that carbon forms bonds with hydrogen in a 1:4 ratio and with oxygen in a 1:2 ratio.

It forms compounds methane CH_4 with hydrogen and carbon dioxide CO_2 with oxygen.

From Mendeleev's Periodic table, we can understand the basic reactivity of carbon.

Carbon belongs to group-14 (according to the modern periodic table) and period-2. Its atomic number is 6, and its atomic mass is 12 (a highly available isotope).

Carbon has many isotopes, but first, what is "Isotope"?

Isotopes are elements with the same atomic number but a different mass number. Carbon has 15 isotopes from ^8C to ^{22}C among which ^{12}C is the most abundant and stable.

Carbon has tetra-valence; that is, carbon has 4- valence electrons for bond formation.

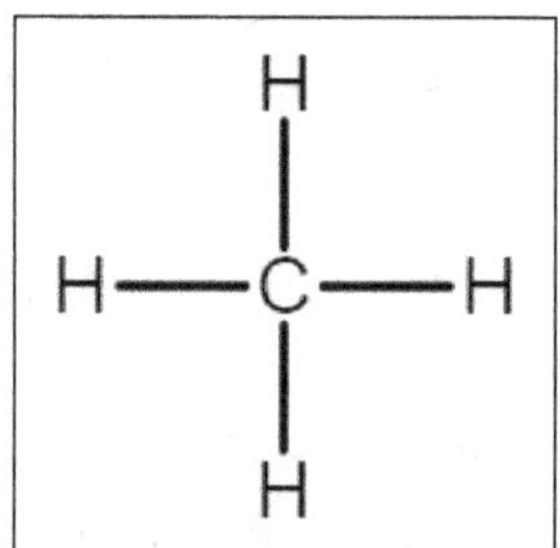

But how does carbon have 4 valence electrons with electronic configuration as $1s^2\,2s^2\,2p^2$?

To obtain octet configuration,the excitation of electron from 2s to 2p orbital there by forming four valence electrons for bonding. Carbons excited electronic configuration is $1s^2\,2s^1\,2p^3$.

Carbon also has a property known as catenation by which it forms bonds with other carbon atoms in the form of series to give long chains.

As the number of carbon increases in the carbon chains, they are known by different Root words:

Alkane table

S.no	Number of carbon	Root word used
1.	One carbon (C)	Meth
2.	Two carbon (C_2)	Eth
3.	Three carbon (C_3)	Prop
4.	Four carbon (C_4)	But
5.	Five carbon (C_5)	Pent
6.	Six carbon (C_6)	Hex
7.	Seven carbon (C_7)	Hept
8.	Eight carbon (C_8)	Oct
9.	Nine carbon (C_9)	Non
10.	Ten carbon (C_{10})	Dec

Chapter 2

How Do We Name Such Compounds?

Organic Chemistry is a vast branch to study, yet there is a lot more awaited to be brought into the limelight. How does someone describe or communicate about an element which can form these many compounds?

That is when a theme of nomenclature is introduced, which deals with naming each and every compound of carbon.

In proper words, nomenclature of organic compounds is the study of naming carbon molecules by following the rules set up by IUPAC (International Union of Pure and Applied Chemistry).

Before we start discussing the rules and priorities of the Functional groups, we should know the suffixes and prefixes of the function groups we use:-

Types of bonds		
Sl.No	Number of bonds between carbons	Suffix used
1.	C——C	-ane Ex:- Methane (CH_4), ethane (C_2H_6) (One sigma bond) single bond.
2.	C══C	-ene Ex:- Ethene (C_2H_4) , Propene (C_3H_6) (One sigma and one pi bond) double bond.

3.	$C \equiv C$	-yne Ex:- Ethyne (C_2H_2), Propyne (C_3H_4) (One sigma and two pi bonds) triple bond.

- These bond types are known as Primary suffixes.
- Priority order in Bonds is Double bond> triple bond > single bond.

But in some cases, the priority order is neglected to satisfy least some rules, which will be discussed in the upcoming pages.

How do we use these for naming a carbon compound?

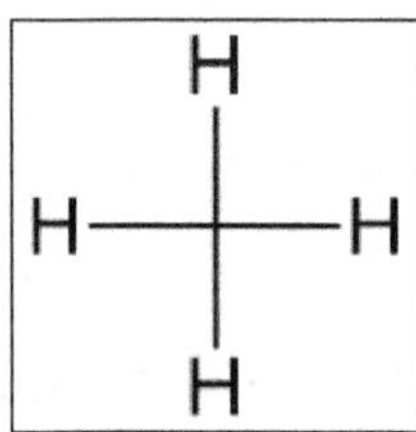

Observe this image; what do you think we will name it? Correct, it is Methane. Very easy, right? Why am I asking such an easy question? But how did this become that easy? We know that there is one carbon and four hydrogen atoms. In that case, we have discussed carbon as "Meth", and in the case of a single bond, it is "ane". So the end compound is Methane.

In the same way, let's try some more compounds. In this compound, there is a carbon-carbon bond and 6 carbon-hydrogen bonds. Hence, it is known as Ethane.

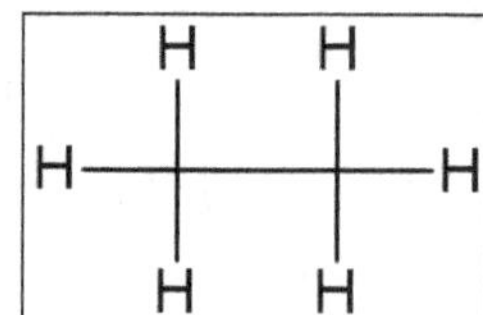

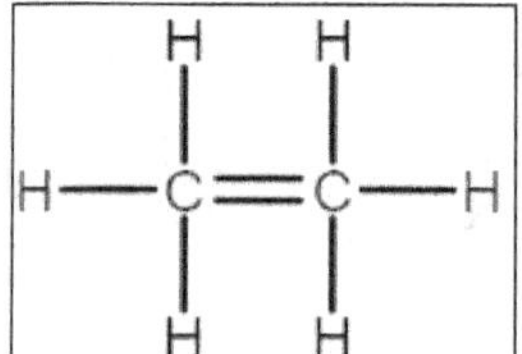

What is the name of this compound? Is it Ethene, as there is a carbon-carbon double bond? But if you think it is Ethene, you are wrong.

If you observe closely, you will see that the two carbons have five bonds, which is not technically possible.

On the other hand, this is the correct structure of "Ethene". All the carbons in this compound have four bonds which satisfy the octet rule. (The bond angle in the structure shall be discussed in the geometry of organic compounds) 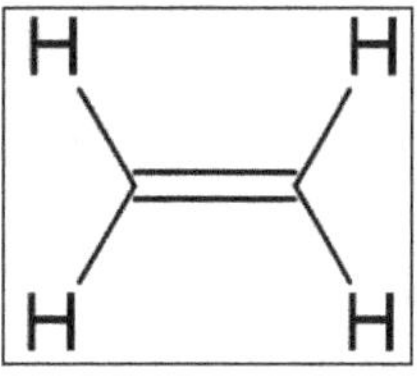

This is the structure of Ethyne or also known as acetylene. The bond between carbon and carbon is a triple bond.

$$H—C≡C—H$$

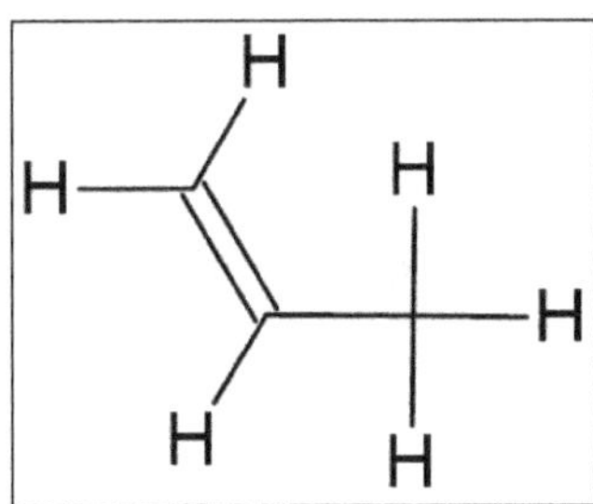

Let's jump a little and discuss Propene and Propyne.

In this image, we can observe that the centre carbon has a double bond with one carbon and a single bond with other carbon. So it is called Propene.

(The double bond is not in the same plane as that of other carbon, this is part of organic geometry)

Here we can observe that the centre carbon has a triple bond with one carbon and a single bond with another carbon. So, the name of the compound is called Propyne. 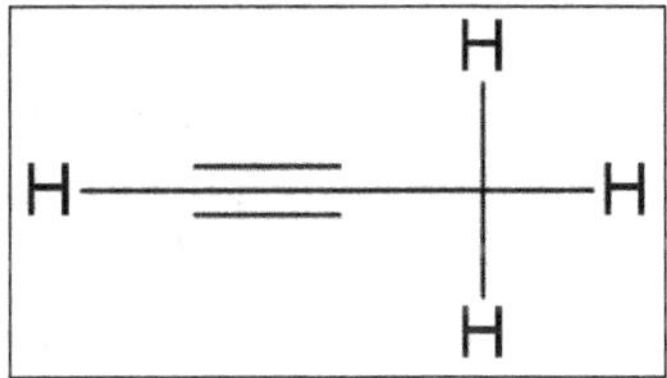

These are the name and structures of some basic compounds.

Functional Groups

These functional groups help in further classification of the compounds and are hence also known as **secondary suffixes**.

Structure of group/formula	Name of group	Suffix used
-OH	<u>Alcohol</u>	ol
-CHO	<u>Aldehyde</u> Carbon has a double bond with oxygen and one valence.	al
-CO-	<u>Ketone</u> Carbon has a double bond with oxygen and double valence.	one
-COOH	<u>Carboxylic acid</u> Carbon has a double bond with one of the oxygen, single bond with OH group and one valence.	Oic Acid
-COOR	<u>Ester</u> Carbon has a double bond with one of the oxygen, single bond with OR group and one valence. (where R- is the alkyl group)	Oate
-COCl	<u>Acid Chloride</u> Carbon has a double bond with one of the oxygen, single bond with Cl and one valence.	Oyl chloride
$-CONH_2$	<u>Amide</u> Carbon has a double bond with one of the oxygen, single bond with NH_2 group and one valance.	amide
-CN	<u>Cyanide</u> Carbon forms a triple bond with Nitrogen and one valence.	Nitrile

-NC	<u>Iso Cyanide</u> In this structure, Nitrogen will have an extra valence after forming bonds with carbon.	Carbylamine or Isonitrile
$-NH_2$	<u>Amine</u> Nitrogen with two bonded hydrogen and one valence.	Amine

Before we understand the pattern of naming compounds, we will have to learn some prefixes.

Prefixes

In general, they indicate the names of substituents and the side chains.

S.No	FORMULA	PREFIX
1.	-F	Fluoro
2.	-Cl	Chloro
3.	-Br	Bromo
4.	-I	Iodo
5.	$-NO_2$	Nitro
6.	$-NH_2$	Amino
7.	-CN	Cyno
8.	-NC	Isocyno
9.	-OH	Hydroxy
10.	-OR	Alkoxy
11.	-NO	Nitroso
These are the Side chain group/ Side Branches		
12.	$-CH_3$	Methyl

13.	$-C_2H_5$	Ethyl
14.	$-C_3H_7$	Propyl

These are a few of the side chains that are commonly used. There are many other side chains that can be created in complex branches. This is a topic to be discussed.

Chapter 3

What Kind of Pattern Do You Think We Observe?

When we talk about the pattern, we mean the repeating functional groups, multiple bonds or the molecule itself (polymers).

But the naming of these compounds has a specific pattern, a particular order of placing functional and main compounds. Now, let's see what the order is and how to use it:-

Prefix + Rootword + Primary Suffix + Second Suffix

Before we start following this pattern for naming compounds, we should know what these terms actually mean.

Prefix: – Terms used to represent side chain attached to the parent carbon chain. It can be a small carbon chain or a functional group.

Root word: – Terms which represent the number of carbons in the parental chain.

Primary suffix: – Term which says what kind of bond is present between the carbon atoms.

Second suffix: – Terms which represent a functional group in the parental chain.

Let's discuss how do we use this form in naming a compound:-

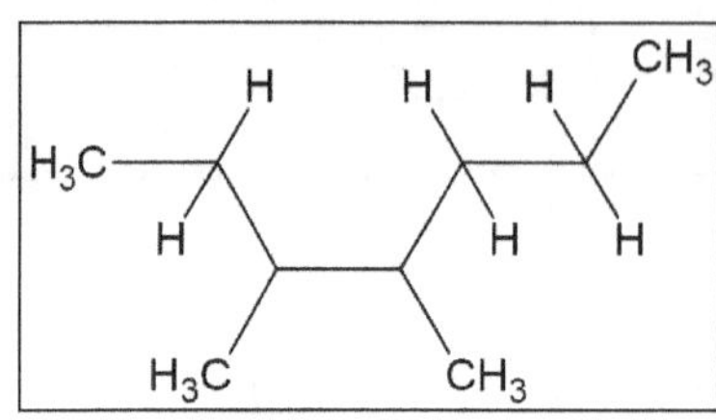

How do we name this compound? Well, you might have started to count the number of carbons. You come to the conclusion that there are 9 carbons, but this compound is not Nonane. So, what should we do now?

Let's consider the longest carbon chain possible in this compound, starting from the top left as the first carbon. If we consider the bottom right, we have 5 carbons in the chain and only 4 if we take the bottom left. Now count to the top right; we have 7 carbons in the chain with two side chains.

Now let's name it,

Prefix:- there are two methyl groups, so we use the term "Dimethyl", and for three, we use Tri, 4-tetra, 5-pent, 6-hexa, 7-septa.

Root word:- Hept because of 7 carbons

The primary suffix is ane since there are no double or triple bonds, and there is no secondary suffix, so the name of the is

3, 4-Dimethylheptane. (Pronounced as 3,4 dimethyl Hept ane)

Now let's take another compound:-

How do we name this compound? Ok, like the previous one, we will start from the top left and take it to the top right. We have the longest chain possible in this carbon (6-carbon atoms).

Then,

Prefixes: – there are two methyl branches. So Dimethyl is one of the side chains. On the top right, carbon has fluorine and the second from the top left has bromine as its functional group.

Root word: – It is Hex since there are six carbons in the parent chain. The primary suffix is 'ene' since there is a double bond. There is no secondary suffix present. So the name of the compound is "2-Bromo-3, 4-Dimethyl-6-fluoro Hexene" is this correct?

Or is the correct among the following

A. 5-Bromo-3, 4-dimethyl-1-fluorohexene

B. 2-Bromo-6-fluoro-3, 4-dimethylhexene

C. **5-Bromo-1-fluoro-3, 4-dimethylhexene**

D. 6-Fluoro-2-bromo-3, 4-dimethylhexene

E. 1-Fluoro-5-bromo-3, 4-dimethylhexene

Which one do you think is correct, and if correct, then why?

Before we discuss the solution for the previous compound. I would like you to name another compound:

How do we name this group? This compound contains two functional groups. Those are the alcohol and carboxylic acid group.

Let us start naming the compound now:-

Prefixes: – As we have seen in the previous table –OH can be considered both as a prefix (Hydroxy) and suffix (Alcohol).

Root word: – The parent chain contains 6 carbon atoms. So its root word is "Hex".

Suffix: – It is a carboxylic acid group (-COOH). It is the suffix in this compound.

In this, why we have considered –OH as prefix and –COOH as suffix will be discussed further during the explanation of the compounds.

So the name of the compound is **3-Hydroxyhexanoic acid**. (3 hydroxy hex an oic acid)

Some of you might wonder how I came to the conclusion of naming this compound. You might be expecting some rocket science law's to name these compounds, but it is not that situation. Rather, it so simple as following "road rules", I compared nomenclature rules with road rules because it is also not possible to reach our destination following all the rules, so there are some "break the rule situation" but not at every point.

Chapter 4

Rules of Naming an Organic Compound

RULE 1:-

Consider the longest chain of carbons as the parent chain; this will determine the root word of the compound.

The rest of the side branches carbons will not be considered while giving the root word, but they will act as the prefixes of the compound name.

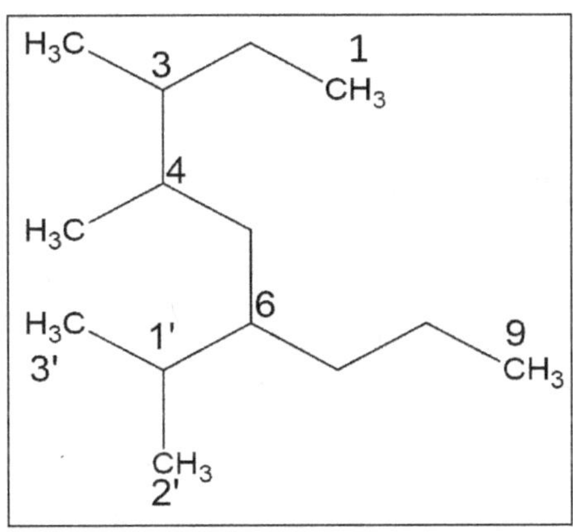

In this image, if you observe closely, there are 9 carbons in the parent chain. So the root word for this compound is "Non".

And further, if we see the third and fourth carbon from carbon-1, those two carbons have methyl branches.

On the carbon-6, there is the complex group which is (1-MethylEthyl). These are the observation from this image.

But you all must be wondering why I have taken carbon-1 as first and not as the last carbon. Any significance in doing that, or was it just a random assumption?

To answer this question will have to learn the second rule of the nomenclature. This rule is known as "The numbering rule" or "The lowest locate rule".

RULE 2:-

While naming an organic compound, we will have to specify the position of each and every branched chain and functional group, including the type of the bond with respect to the parent or the main chain.

In this process, for ease of naming, we follow the method called "The lowest sum rule" or "The lowest locate rule". In this method numbering of carbons is done in such a way that the sum of all the tokened carbons with any kind of functional group must be the least.

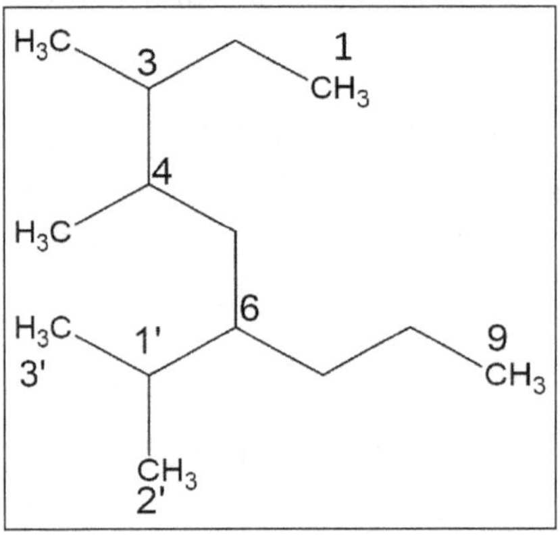

If we observe closely, there are 3 branches in this compound; from carbon-1, the branches are at 3, 4, and 6 positions and with respect to carbon-9, the positions are 4, 6, and 7.

After observing the positions, we can satisfy the lowest sum rule only with respect to carbon-1.

We know the parent chain in this molecule. So start numbering the compound from top left to right and vice versa. Then find the sum of the functional groups in either case. Now we will easily be able to number the functional groups in this carbon.

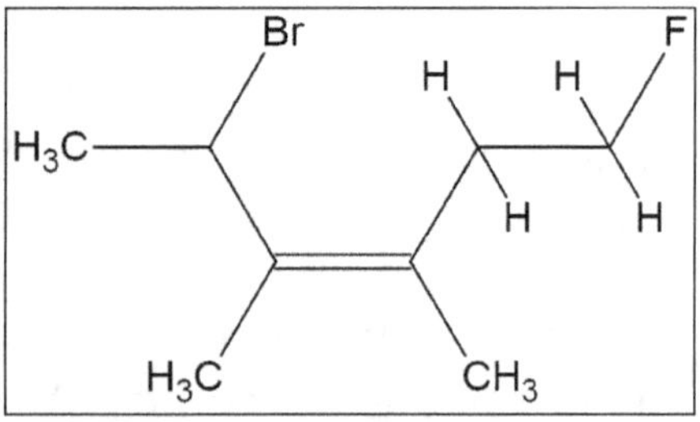

Will that be enough to name the above compound? What else do you think is needed to name them perfectly?

If you are thinking about the order of the functional groups, then you are right. True, after numbering the compound, we actually have no idea which functional group should be named first.

How to use multiplicity in normal branches and multiplicity in complex branches?

RULE 3:-

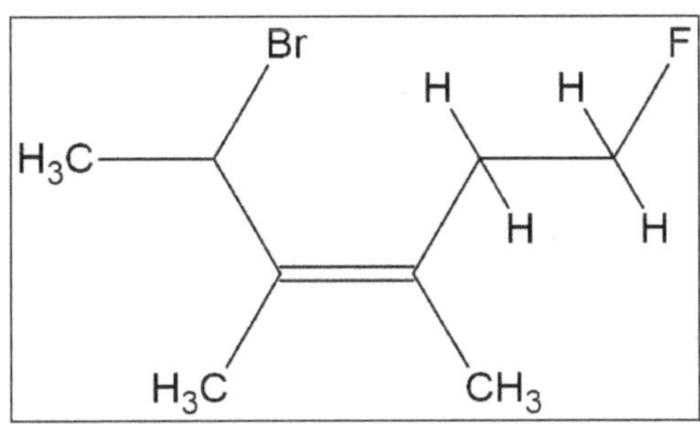

This is the rule exclusively to be followed while righting the name of the compound. The rule says, "the prefixed should be arranged following the alphabetical order and their position being mentioned preceding them."

Now, I guess we will be able to name this compound. We already numbered this compound, we know which carbon should be first, and we can find the position of each prefix from the numbering of the chain.

From the above discussion, we can conclude that name of the compound is

5-Bromo-1-fluoro-3,4-dimethylhex-3-ene. In this compound, after following the alphabetical order, we have placed bromine first, after which we have fluorine followed by dimethyl, but you must be concerned that "D" comes before the "F", then why did I write fluoro before dimethyl.

That is when the concept of "Multiplicity" is introduced.

You must have seen compounds which have two or more same branches or functional groups. In that case, we use multiplicity terms to express all the prefixes instead of writing them each time.

Simple Multiplicity:-

When a functional or side chain group repeats 1) 2 times – Di 2) 3 times – Tri 3) 4 times - tetra 4) 5 times – pent 5) 6 times – hexa 6) 7 times – septa

In the case of Simple Multiplicity, terms given as a prefix to the functional group will not be considered while checking the alphabetical order. Like ethyl will be placed before dimethyl will write the name of the compound.

Similarly, Methyl will be placed before Propyl

Dichloro will be placed before fluoro etc.

Complex Multiplicity:-

Previously I have used terms such as complex branch; let's discuss that in detail.

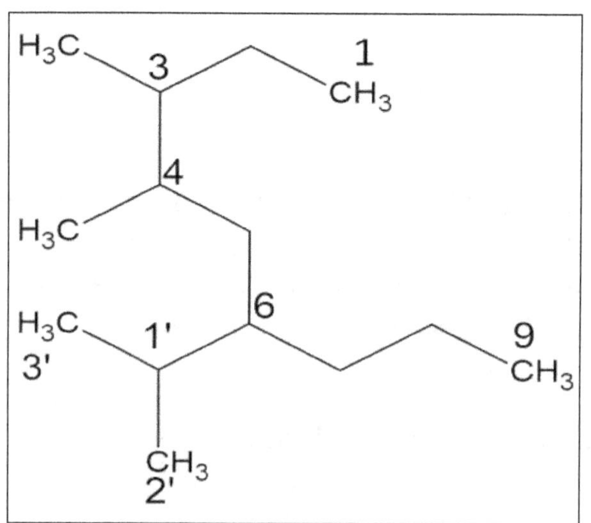

In this picture, there is a complex branch in the 6th position. Consider 'R' to be the main chain; count the number of carbon in the side chain, that is, 3 carbons with centre carbon attached to the main chain. So, in the side chain, there is a chain with 1' & 2' carbons as the main chain and a branch with 3'-carbon. If we had written its name, it would be 1-Methylethane, but since this whole compound is a side branch for R, the name of the compound will be (1-Methylethyl).

In this way will have to represent complex branches in brackets while writing a compound's name.

The nomenclature used while representing two or more of the same complex group is as follows:

(1) 2 times – Bis (2) 3 times – Tris (3) 4 times – Tetrakis (4) 5 times – Pentakis (5) 6 times – Hexakis (6) 7 times – Septakis

Now let's name the compound mentioned above:-

6-(1-Methylethyl)-3,4-dimethyl-nonane

So far, we have started to learn how to name a compound, we still have a long way to go for completion, but before that, we have to discuss the representation or drawing of the organic compound.

Until now, we have mentioned every carbon present in the compound, which is time-consuming and hard to understand when there is a complex structure.

That is why the introduction of "Bond Line structures" was done.

Chapter 5

Bond Line Structures

As the name suggests, this representation consists of lines considered bonds, where every intersection or terminal represents carbon, and the line represents the bond between the two carbons.

In the case of any functional groups, they are mentioned in their condensed form like –COOH, -CO-, -F, -Br... etc.

Example of C_3H_8

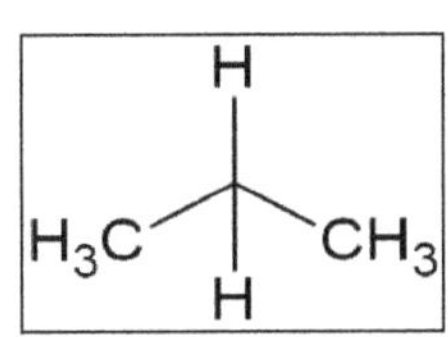

If we apply the dot line structure to that, we will get this structure.

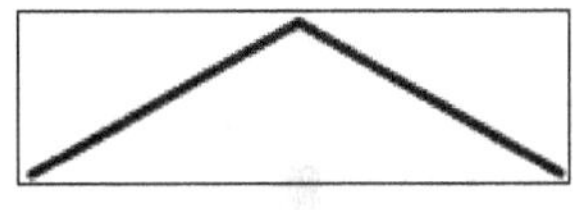

Similarly, when we can draw the structure of **2-Bromo-1,4-difluoro-3-methylbut-2ene** as follow:-

The structure of the compound 5-Amino-3-Bromopentanal is as follows:-

<u>Bird Out of the Box:-</u> Try representing the following compounds in Bond line structures:-

A.   3-Bromo-1-isocyno-5-methylhexane

B.   4-iodo-6-methyloct-1-yne

C.   Hex-3-ynoic acid

Try naming the following compounds given in Bond line structures:-

(a)

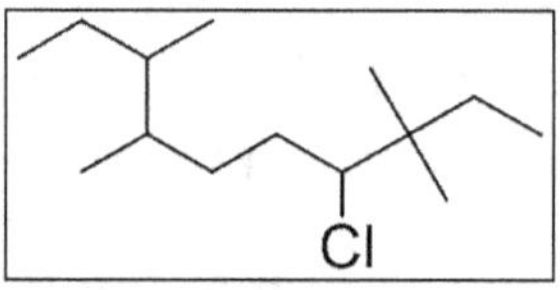

(b)

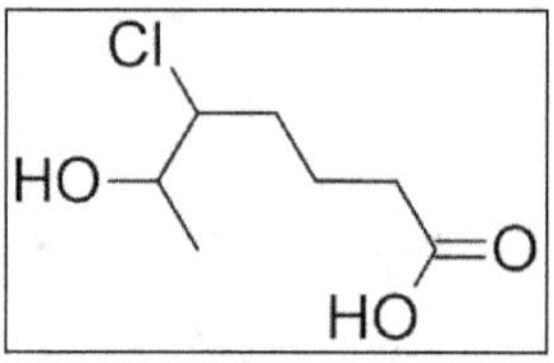

Chapter 6

Mono-Functional Groups

Previously while discussing a compound named **3-Hydroxy Hexanoic acid,** I have told that "-OH" has both suffix and a prefix, whereas "-COOH" has only a suffix. This isn't the accurate reason for using "-COOH" as the primary functional group.

While selecting a parent chain and numbering it, there is a priority order which has to be followed.

The order to be followed is: first, consider a chain with a Functional group; secondly, choose carbons with multiple bonds, then select the chain with the largest number of side chains.

Functional Groups > Multiple bonds > Side chains

Let's discuss some **Mono-functional groups/ Chain terminating functional groups**.

1. –COOH - Oic Acid

2. –COOR - Oate

3. –COCl - Oyl Chloride

4. –CONH$_2$ - Amide

5. –CN - Nitrile

6. –CHO - al

While naming a functional group with a chain-terminating functional group, allocate the carbon in the functional group with position 1. We need not mention the position of the chain terminating functional group. According to IUPAC, they are always allocated position 1.

Naming different compounds with Functional groups:-

If we observe closely, the chain terminating functional group is "-COOH". Considering that carbon is first, we will have number the longest chain possible.

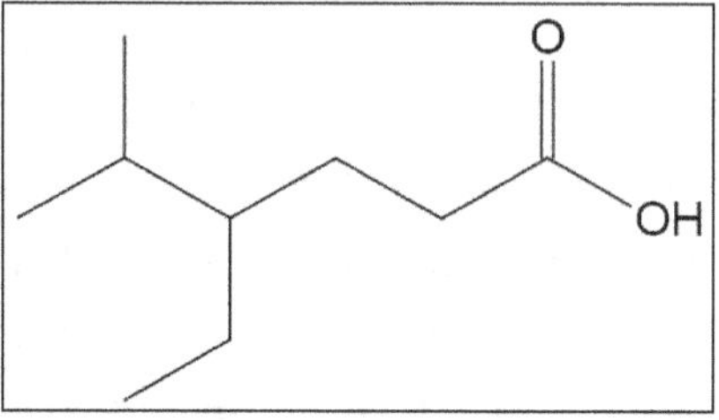

There are two ways possible in which both have 6 carbons. Initially, we discussed considering a functional group, then multiple bonds followed by side chains.

Since there are no multiple bonds in this compound, we will have to look for a chain with more side chains. That is, one of the chains has only one complex group, whereas the other chain has two simple, functional groups.

So, the name of the compound will be **4-Ethyl-5-methylhexanoic acid**.

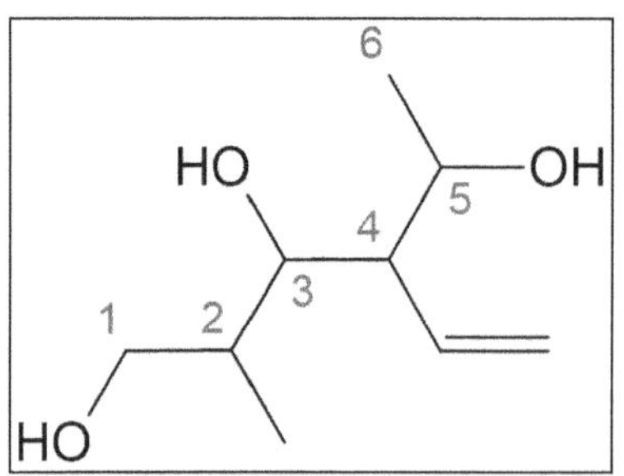

In this compound, there are 3 functional groups in different positions. There is also a multiple bond in this compound. But if we follow the order, the functional group should be given more priority than multiple bonds.

After numbering this compound, we can consider the chain with multiple bond as a side chain. There are 3 functional groups in 1, 3, and 5 positions.

So the two branches, Ethenyl and Methyl, are in positions 4 and 2, respectively.

The name of the compound is **4-Ethenyl-2-methylhexa-1,3,5-triol**.

The functional group in this compound is Ester. The main chain has 5 carbons and an ethyl side chain.

Note:- While writing an Ester compounds name, mention the "R" groups name in the first where R'-COOR, R' is the main chain.

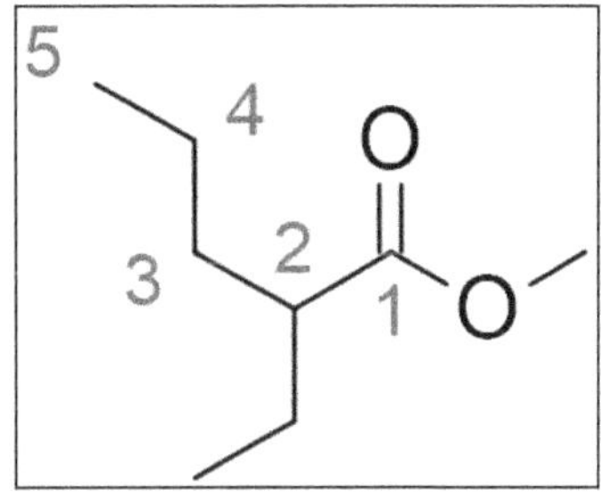

So, in this compound, if taken in R'-COOR form, R is Methyl. In that case, mention Methyl before starting the original chain.

The name goes as follows **Methyl-2-ethylpentanoate**.

Chapter 7

Amides Vs Amines

<table>
<tr><td>

The organic compound with "Amide" functional group has three forms:-

  i. R-CO-NH$_2$ It is 1°-amide

  ii. R-CO-NH-R' It is 2°-amide

  iii. R-CO-NR'-R" It is 3°-amide

When it is used as a secondary suffix it is called an Amide.

When used in a prefix, it is called Carbamoyl or amido.

Naming of Amide:-

In this primary amide, there are 6 carbons in the parent chain. There are no functional groups or side chains other than amide. The name of this compound is **Hexanamide.**

</td><td>

The organic compound with "Amine" functional group has three forms:-

  i. R-NH$_2$ It is 1°-amine

  ii. R-NH-R' It is 2°-amine

  iii. R-NR'-R" It is 3°-amine

When it is used as a secondary suffix it is called an Amine.

When used in a prefix, it is called amino.

Naming of Amine:-

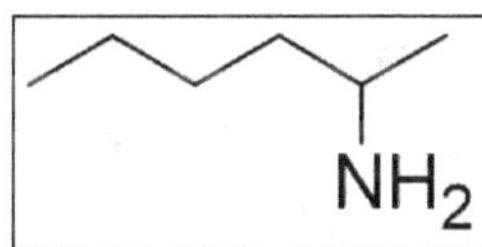

In this primary amine, there are 6 carbons in the parent chain. There are no functional groups or side chains other than amine on the second carbon. The name of this compound is **Hexan-2-amine.**

</td></tr>
</table>

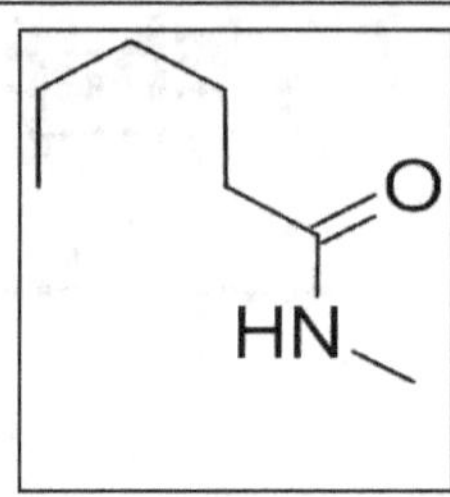

In this secondary amide, the parent chain has 6 carbons. There is methyl attached to Nitrogen. There are no other functional groups or side chains on the parent chain. The name of the compound will be

N-Methylhexanamide.

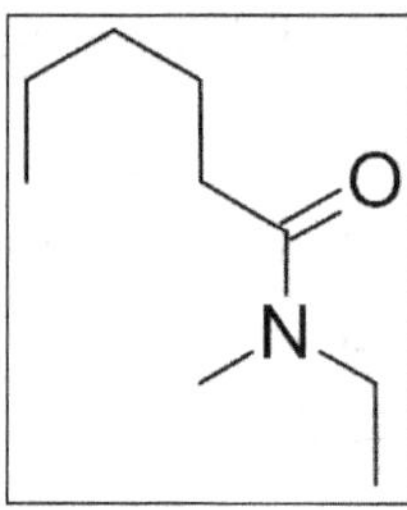

In this tertiary amide, the parent chain has 6 carbons. There is ethyl and methyl attached to Nitrogen. There no other functional groups or side chains on the parent chain. The name of the compound will be

N-Ethyl-N-methylhexanamide.

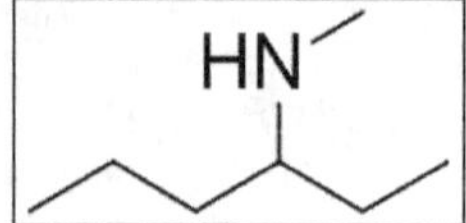

In this secondary amine, there are 6 carbons in the parent chain. There are no functional groups or side chains other than amine on the third carbon. The name of this compound is

N-Methylhexan-3-amine.

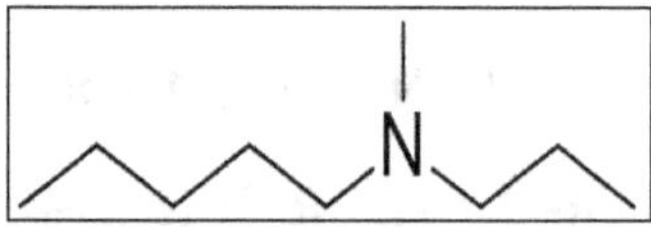

In this tertiary amine, there are 6 carbons in the parent chain. There are no functional groups or side chains other than the amine. The name of this compound is

N-Methyl-N-propylpentanamine.

Chapter 8

Poly-Functional Groups

We have discussed how to name a compound with one functional group. What if? There are more than one functional group to an organic compound. Which functional group's carbon should be considered in the main chain, or which is the main functional group?

You must need an order of considering functional groups. Precisely, you need the priority order of functional groups, which goes as follows

R-COOH> R-SO$_3$H> R-COOR> R-COCl> R-CONH$_2$> R-CN> R-CHO> R-CO-R'> R-OH> R-NH$_2$

A compound with the highest priority will be considered as the main functional group in the given compound. The next question that arises is how do we differentiate priority groups from normal functional groups while naming them?

There has to be some kind of differentiation in the nomenclature of these terms. Let's see how we differentiate them.

Functional groups	Secondary suffix	Prefix	Special suffix
R-COOH	-Oic Acid	Carboxy-	Carboxylic acid
R-COCl	-Oyl Chloride	ChloroCarbonyle	Carbonyl chloride

R-CONH$_2$	-Amide	Carbamoyl	Carboxamide
R-CN	-Nitrile	Cyano	Carbonitrile
R-CHO	-Al	Formyl	Carbaldhyde
R-COOR'	-Oate	Alkoxy Carbonyle	Carboxylate
R-COR'	-One	Oxo	-

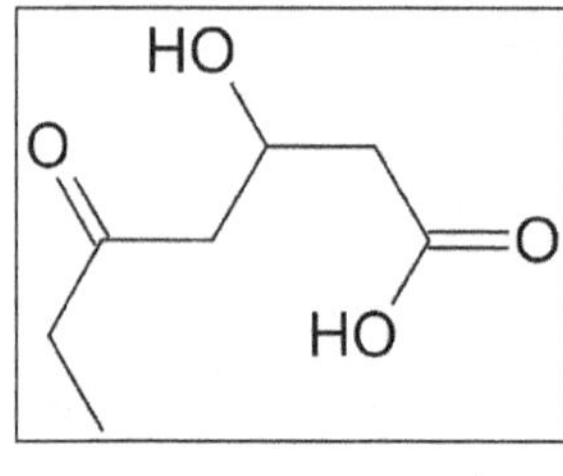

In this compound, if we follow the numbering given, we can Identify 3 functional. After checking with the priority order, we can conclude that –COOH>-CO->-OH. But in this case, carboxylic acid is the main functional group; the rest of the compounds can be considered in prefixes.

Now, considering all the rules and priority order in nomenclature, we can name this compound **3-Hydroxy-5-oxo-heptanoic acid**.

As we observe this compound, we can spot three functions those are amide, amine and aldehyde. According to the priority order, the amide is the main functional group. The name of this compound will be **3-amino-8-oxo-octanamide**.

In this compound, we can see there are two similar functions that are nitrile group and a double bond. When there are 3 or more terminating functional groups, do not consider the functional group in the main chain and use a special suffix. The name of this compound is **Hex-1-ene-1,6,6-tricarbonitrile**.

Before we go into another topic, there are somethings we haven't discussed yet. What are the common names of a few frequently repeating compounds?

The common name of organic compounds:-

Till now, we have been studying how to write the name of a compound. If you remember, we have discussed some of the complex side chains and their multiplicity terms. We have learnt how to name a complex side chain and write the whole compound's name. What if I tell you that there is a simple way to write the name of a complex branch which is accepted by IUPAC? Sorry, Sorry... I can hear raging voices of your thinking; why didn't I mention this earlier? You know there is a saying that "Late than Never", it means it is better to arrive late instead of never reaching.

In this topic, we will discuss some of the complex branches and their common names. Before, we also have to see some of the "Root Word's common names".

C_n (number of carbons in chain)	Common Chain name
C_1	Form
C_2	Acet
C_3	Propion
C_4	Butyr
C_5	Valer

These root words can be used to write the name in a simpler way; see some of the examples below to understand

HCOOH – Formic Acid (1 carbon and carboxylic acid group)

CH_3COOH – Acetic Acid (2 carbon and carboxylic acid group)

CH_3CH_2COOH – Propionic acid(3 carbon and carboxylic acid group)

$CH_3CH_2CH_2COOH$ – Butyric Acid (4 carbon and carboxylic acid group)

CH$_3$CH$_2$CH$_2$CH$_2$COOH – Valeric Acid (5 carbon and carboxylic acid group)

CH$_3$CHO – Acetaldehyde (2 carbon and Aldehyde group)

CH$_3$CH$_2$COCl – Propionyl Chloride (2 carbon and Carbonyl Chloride group)

In the same way, some complex branches and alkyl compounds, branches have common names. These can either be a complex branch of an organic compound, or they can act themselves as a compound depending on their valence.

Structure of the compound/Branch	Actual name (following all nomenclature rules)	Common Name
H$_3$C—Br	Bromomethane	Methyl bromide
	2-Chloropropane	Iso-propyl chloride
	1-Bromobutane	n-butyl bromide
	2-Bromobutane	Sec-butyl bromide

	1-Bromo-2-methylpropane	Iso-butyl bromide
	2-Bromo-2-methylpropane	*Tert*-butyl bromide
	1-Bromo-2,2-dimethylpropane	Neo-pentyl bromide
I have chosen only the bromine functional group to keep it simple, and other functional groups can be used in its place.		

Now we have discussed some common names of the complex compounds, but none of these complex compounds has multiple bonds, and not all of the terminating functional groups are discussed. Now we shall see some of the common names of the compounds with multiple bonds and common names for some ethers and ketones.

Side chain/ Esters/Ketone Any main alkyl group)	Common Name
	Vinyl side chain
	Allyl side chain

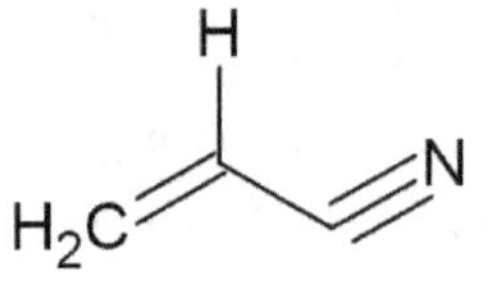	Vinyl Cyanide or Acrylonitrile
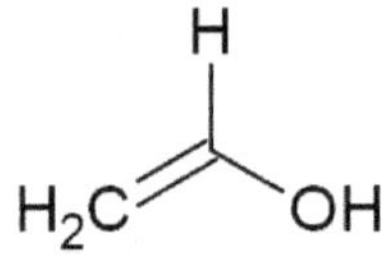	Vinyl Alcohol
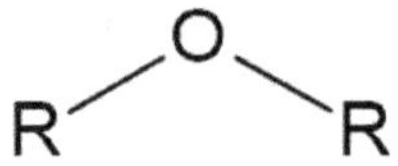	It is symmetrical ether. Dialkyl ether
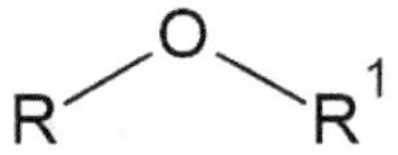	It is unsymmetrical ether. 1-alkyl 2-alkyl ether.
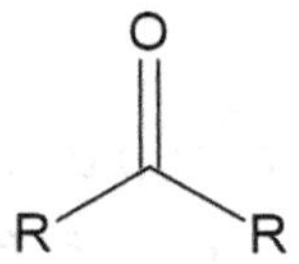	It is a symmetrical ketone. Dialkyl Ketone
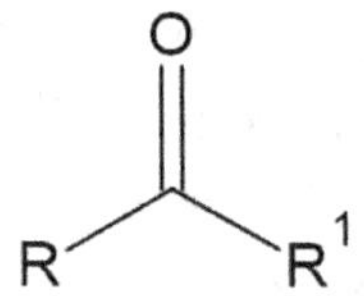	It is an unsymmetrical Ketone. 1-alkyl 2-alkyl Ketone
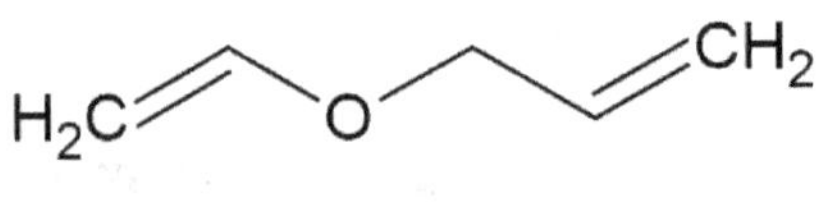	Allyl Vinyl Ether

These are some of the most common compounds which we will be recollecting in the future. These common are terms accepted by IUPAC and can be used in most cases, but in some situations, there will be a need to use their actual name or the IUPAC nomenclature preferable.

Chapter 9

Cyclic Compounds and Its Nomenclature

We have learned about long chain compounds with different functional groups, and we have learnt how to name them. But what if the two end carbons are joined together, forming a closed compound? How are those compounds named, and what properties do they possess?

As the name suggests, these compounds are closed organic compounds with no terminating carbon in the parent chain. Carbon chains with three or more carbons can form cyclic organic compounds. It is similar logic which was taught to you in mathematics while discussing closed figures. There is no closed figure without vertices and lines, but here the vertex is replaced by carbon, and the line represents the bond between them.

Nomenclature of simple cyclic compounds:-

S.No	Structure of the compound	Name of the compound
1.	△	This compound has three carbons, and it is cyclic. So the name of this compound is **Cyclopropane**.
2.	▢	This compound has four carbons, and it is cyclic. So the name of this compound is **Cyclobutane**.

3.		This compound has five carbons, and it is cyclic. So the name of this compound is **Cyclopentane**.
4.		This compound has six carbons, and it is cyclic. So the name of this compound is **Cyclohexane**.
5.		This compound has seven carbons, and it is cyclic. So the name of this compound is **Cycloheptane**.
6.		This compound has eight carbons, and it is cyclic. So the name of this compound is **Cyclooctane**.

In the above compounds, we have used the term "Cyclo", which is used before the root word, which represents the compound as a ring. Now, we will learn some rules to be followed while naming cyclic compounds.

RULE 1:-

When a compound consists of both ring and open chains, selecting the parent chain depends on the number of carbons individually in the ring and open chain. We will have to opt for the one with more compounds.

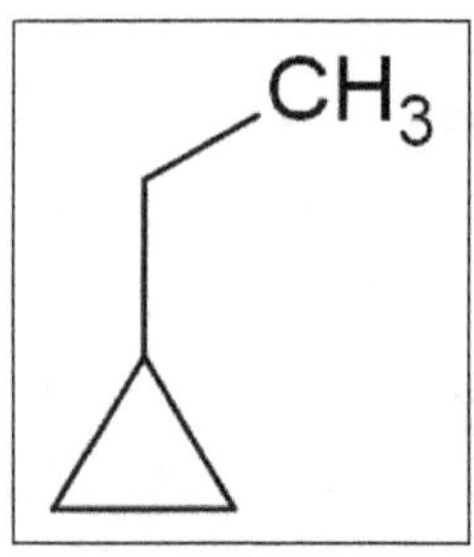

There is a three-carbon ring with two carbon chains. The name of this compound is **Ethyl-cyclopropane**.

There is a five-carbon ring and a six-carbon chain. The name of this compound is

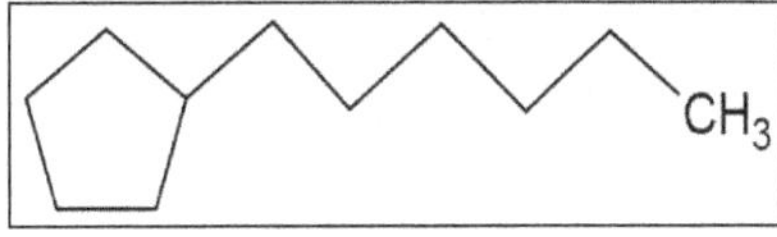

1-Cyclopentylhexane.

What if:-

If the ring and open chain had the same number of compounds, which one should we consider as the parent chain?

1. If there are any functional groups on either of them, consider the one which has more priority as a functional group.

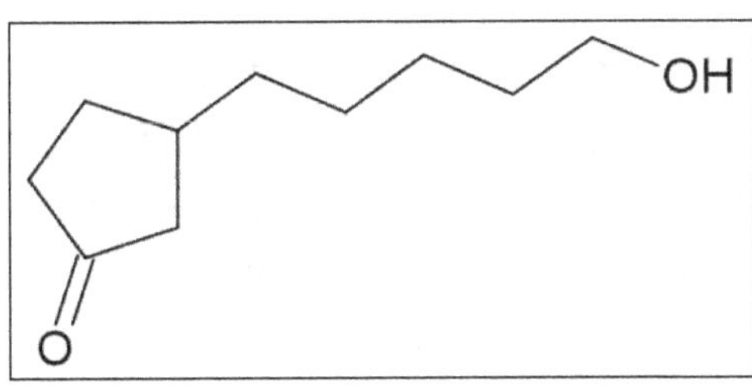

In this compound, we can observe that both the ring and open chain have the same number of carbons. So we can observe two different functional groups on each of them. After checking with the priority order, we will have to consider the ring with the ketone group. So the name of the compound is **3-(5-Hydroxypentyl)-cyclopentanone**.

2. The open chain has less number of carbons but has more priority groups. How do we name such a situation?

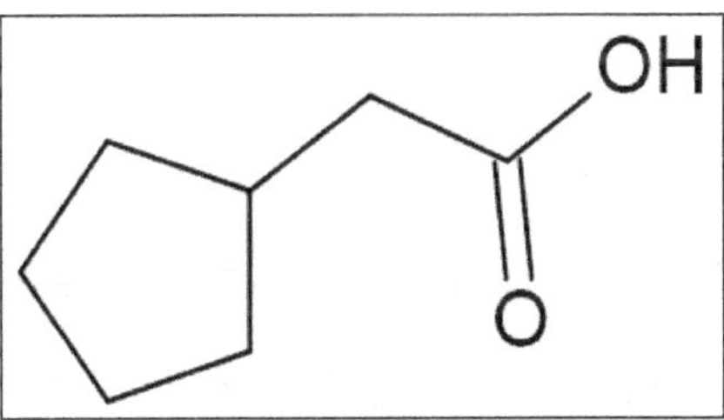

In this compound, we can see that the open chain has only two carbons, but carboxylic acid acts as its functional group. In this, we will consider the open chain as the parent and name the compound. So the name of this compound will be

1-(Cyclopentyl)ethanoic Acid (or) **1-(Cyclopentyl)ethane carboxylic acid**.

3. If the chain contains two cyclic compounds, how do we name them in such a situation?

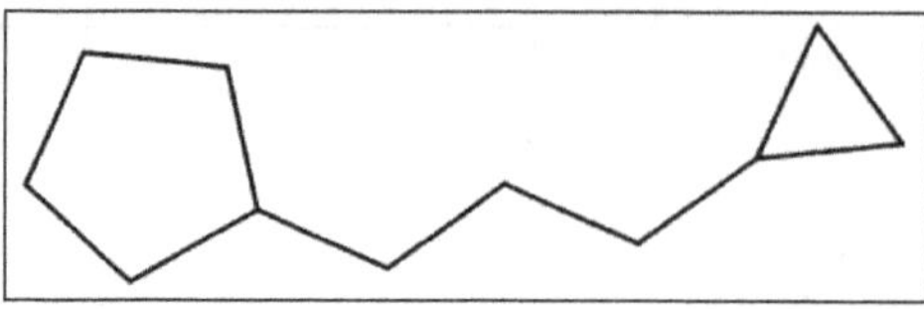

In this compound, there are two ring structures attached to a small propane chain at terminating points. In such scenarios, consider the ring with more carboons is the parent chain.

The name of this compound is (**3-Cyclopropylpropyl)-cyclopentane**.

RULE 2:-

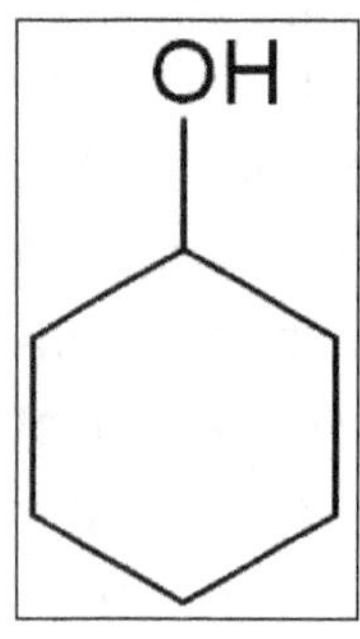

Naming cyclic compounds with different functional groups, multiple bonds and side chains. We will be following the same rules as we have done for open chain compound that is priority orders will not change while naming cyclic compounds.

If we name this compound which has one functional group, it will be **Cyclohexanol**.

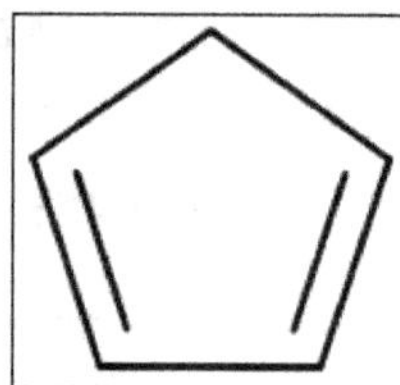

The name of this compound with two double bonds will be **Cyclopent-1,3-diene**.

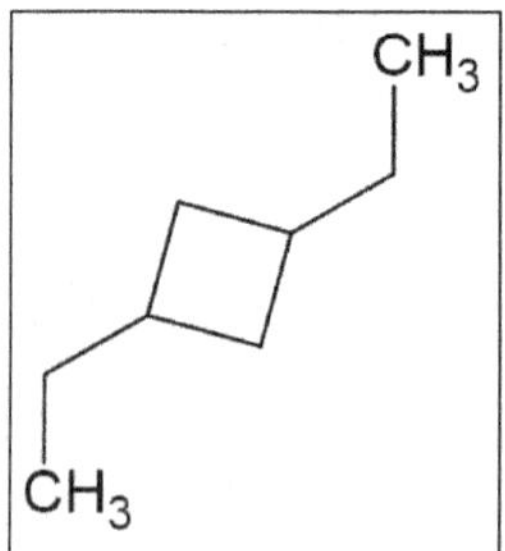

The name of this compound with two side chains will be

1,3-Diethylcyclobutane.

We have learnt all the rule. Can you try naming this compound? I will lend you some leads: functional groups have more priority over multiple bonds, and chains with

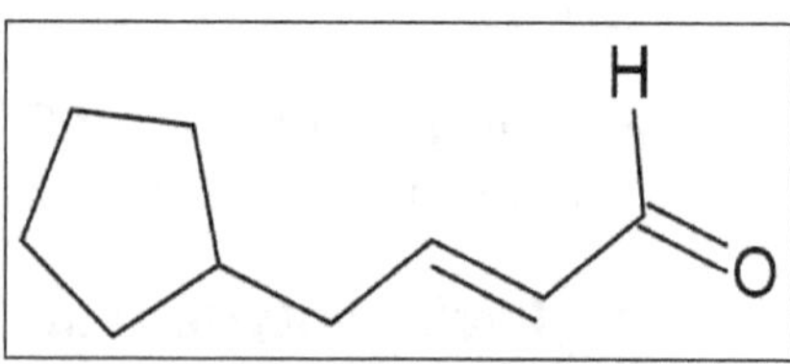

terminating functional groups are given more importance.

If you have named it as 3-CycloPentyl-Propenal, you are incorrect. Because the position of the double bond is not mentioned, that is **3-Cyclopentylprop-3-en-1-al**.

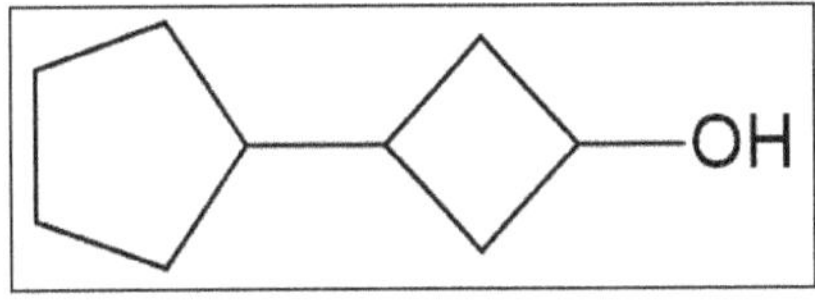 In this compound, we will have to consider the chain or ring with the functional group as the parent carbon chain. So the name of this compound will be **3-Cyclopentyl-cyclobutane-1-ol.**

In this compound, there are two functional groups that are repeating. Check which has more priority. It has two Carbonitrile and two ketone groups. In which the nitrile group has more priority than the ketone group. 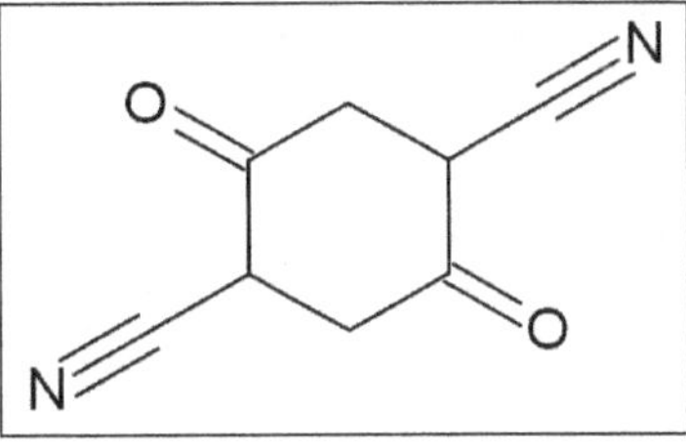 So the name of this compound is **2, 5-Dioxo-cyclohexane-1, 4-dicarbonitrile**.

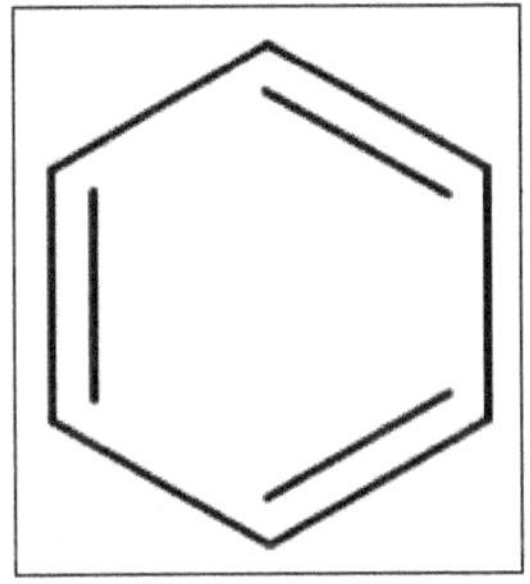 What is the compound name of this compound? If you are thinking of 1,3,5-TricycloHexane, you are not wrong, but for this compound, based on its properties, different compounds it forms, and its stability, it is called **Benzene**.

Why is a special name to be given to such a simple compound?

What are the different compounds it forms?

What are the properties I mentioned while using the word "Benzene"?

How is Benzene stable, and what properties cause this high stability?

All these will be answered and discussed elaborately in the next topic called Aromaticity or Aromatic compounds, in which we will deal with resonance, hyperconjugation, and other topic related to the stability of the compound and differentiating aromatic compounds from other compounds.

Aromaticity

Before we start with the aromatic nature of the compounds, we should learn about the resonance and resonating structures. What is resonance, and what is its use?

Resonance is a way of representing the bonding in certain molecules and polyatomic ions based on hybrid structures in valence bond theory. It is a way of analyzing delocalized electrons in case they were not explained by one single Lewis structure.

The above structure represents the different resonance structures of the ion CO_3^-.

Why is there a change in the position of charge or electrons in such ions or compounds? When there is more than one way to place double bonds and lone pairs on atoms, that is when resonance structures arise. There is a constant shift in electron pairs in such atoms. This shifting is known as delocalization of electrons. Due to this delocalization of electrons, the molecules obtain higher stability.

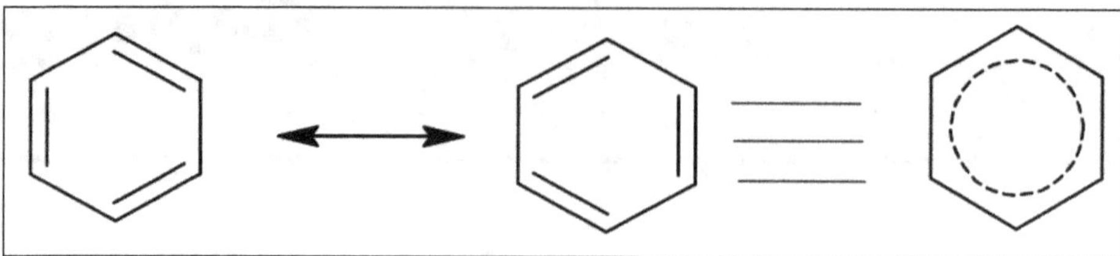

In general, resonating structures only differ in the placement of their pi-electrons, which means the original structure (sigma bond skeleton structure) does not undergo any changes.

The obtained structures after resonance will have the same number of paired electrons and unpaired electrons (radical) as that of the original structure. In the presence of any element with lone pair of electrons, in such cases, only one lone pair will participate in resonance structures.

From the following example, we can observe the change in the position of lone pair electrons and the balance of the charges to maintain the overall neutrality of the compound.

Among all the resonating structures, to predict the most stable one should follow these given conditions:

> Neutral structures are more stable than charged structures.

> Structures with more covalent bonds are more stable.

> Compounds with a complete octet on every atom are more stable, even if an electronegative atom has a positive charge on them.

> In the case of charge separation, the compound with an electronegative atom having a negative charge is more stable than compounds with an electronegative atom having a positive charge.

> So far, we have discussed resonance and resonating structures, the importance of such structures, notes before representing a resonating structure, the charge neutrality in such structures and different examples.

> Now, we can enter into the topic of aromatic compounds and discuss their properties and daily uses.

From the name aromatic compounds, we can conclude they have something related to sniffing. Yes, the word aromatic is given because these hydrocarbons have a distinct pleasant aroma. One of the best examples to experience the pleasant smell of aromatic compounds is Naphthalene Balls.

The chemical name is Naphthalene, and its structure is

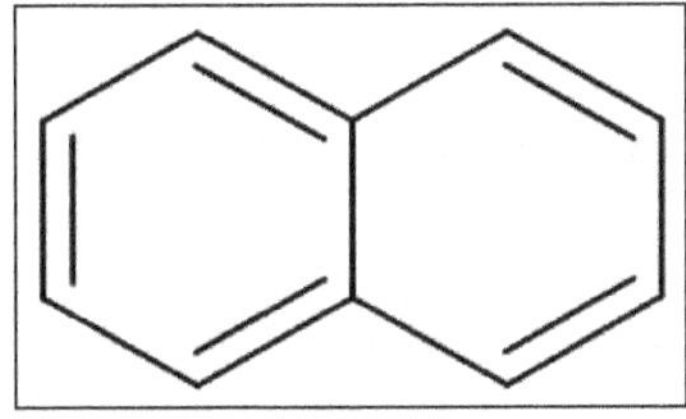

Naphthalene

How do we identify an aromatic compound from other compounds?

It was in Germany year 1931 that a man named Erich Huckel, who was an expert in both physics and chemistry, proposed a theory that would help in differentiating aromatic compounds from others based on Molecular Orbital Theory.

Huckel's Molecular Orbital theory or Huckel's approximation is used to determine the energies and the shape of the π molecular orbital. Through this, we will be able to study and understand the arrangement of **π molecular orbital. The limitation** of this model is that it only addresses conjugated hydrocarbons, and only **π electron molecular orbitals are included**; σ electrons are not considered or neglected.

HUCKEL'S RULES to identify if the given compound is an aromatic compound:-

1. The compound should be a cyclic compound.

2. All the elements in the ring must be in the same plane, or the compound must be planar.

3. The number of electrons which are being delocalized must be in $4n+2$ in the form of a π bond or as a lone pair of electrons (n value can be from 0 to infinite).

4. All the delocalized electrons should be in conjugation.

How will we classify a compound which does not follow any one of Huckel's Rules?

We have some the classification which can be considered given below:-

S. No	Properties	Aromatic Compounds	Anti-Aromatic Compounds	Non-Aromatic Compounds
1.	Shape	Cyclic and Planar	Cyclic and Planar	Will fail minimum of one condition.
2.	Number of Electrons	$4n+2$ delocalized electrons.	$4n$ delocalized electrons	
3.	Stability	Highly Stable	Highly Unstable	Unstable

Aromatic compounds are divided into two variations which are:-

i.   Benzenoids: compounds with one benzene ring.

ii. Non-Benzenoids: compounds without benzene ring.

Now, let us get familiar with some of the Benzenoids:-

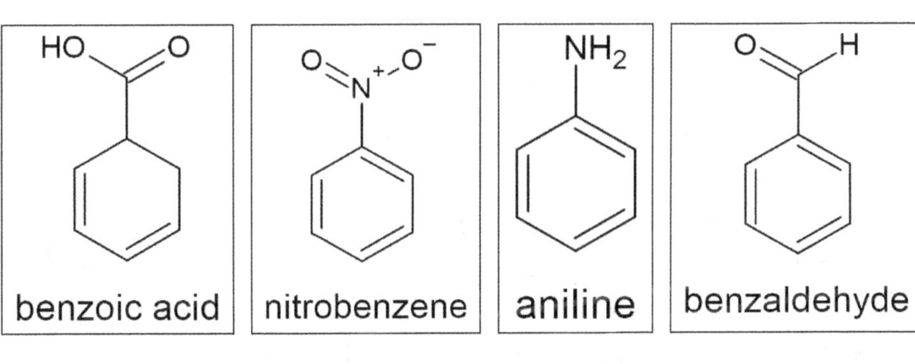

(or) Acetophenone

These are some of the Benzenoids; there are multiple uses of these benzene derivatives. They are used in pharmaceuticals, and many drugs and tranquillizers contain these compounds. They are used in making pipes, parts for automobiles etc.

Why don't we discuss this usage in detail?

We all have used the medicine paracetamol, and it is generally used for aches and as a pain killer; sometimes, we even use it to bring down temperature. It is often used as a remedy for treating colds and flu. If you look at the structure, you will understand the importance of the benzenoids.

N-(4-hydroxyphenyl)acetamide

The IUPAC name given to paracetamol is N-(4-hydroxyphenyl) acetamide. If you remember the rule, we have discussed ketone has a higher priority as a functional group than alcohol. So, in this situation, we use acetate as the parent chain, and we also use the term acetate instead of ethyl because IUPAC accepts some of the root words being replaced by their common names.

Now, let us discuss something interesting. Imagine yourself travelling in a car weighing 75Kg which is moving at a speed 12Kmph and has a deceleration of $2km/h^2$, and finally comes to zero. These are the questions you are asked in physics: you then be asked to calculate the distance travelled or force applied to stop the vehicles. But have you ever wondered what the tire is made up of?

Well, to speak frankly, the tire is made of natural rubber, synthetic rubber, metals and other materials like silica, nylon and etc. But we will only discuss the synthetic fibre in specific.

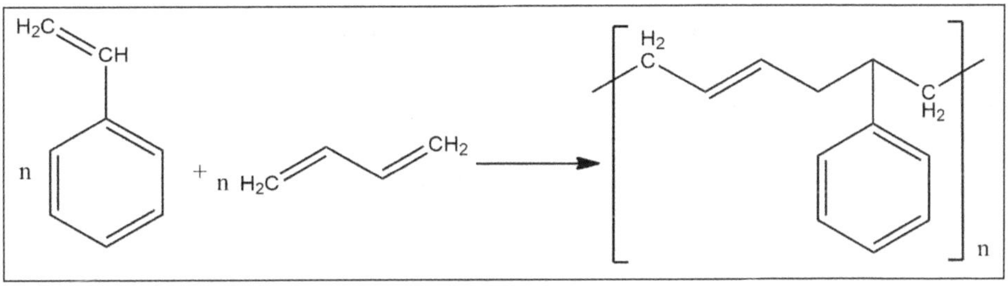

Styrene **But-1,3-diene** **Buna-s**

Ethenylbenzene is the IUPAC name given to styrene, and this is one place where we use benzenoids.

What are the Non-Benzenoids? Before we discuss these compounds, I would like to take you all on a tour of Replacement Nomenclature.

Replacement Nomenclature:-

This is just a way of subdivision of common naming a compound; in these, most of the carbons in a compound are replaced by inorganic elements like Nitrogen (N), oxygen (O), and sulphur (S).

Let us take some examples to understand the meaning of the above sentence.

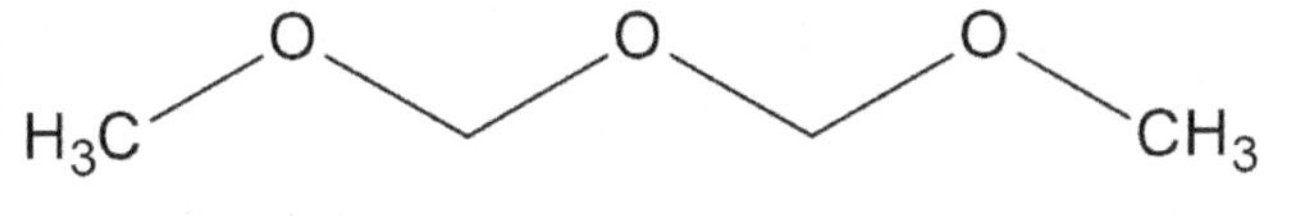

methoxy(methoxymethoxy)methane

The common name of this compound is **2,4,6-trioxaheptane.** In this compound, you can observe ease in naming the compound using common names and replacement terms.

So the function of the replacement term is to help name tough compounds using common names. How many replacement terms exist?

Well, we will be discussing the replacement terms for Nitrogen, oxygen and sulphur.

S.No	Element	Replacement Prefix
1.	Nitrogen	Aza
2.	Oxygen	Oxa
3.	Sulphur	Thia

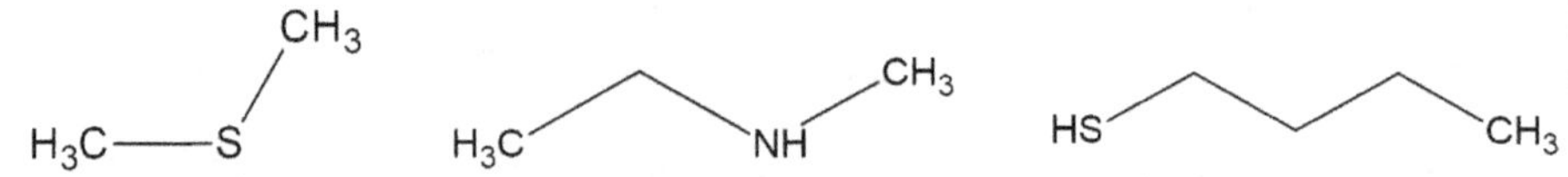

Dimethyl sulfide   N-methylethanamine   butane-1-thiol

The common names or the names given after using replacement terms are

2-Thiapropane, 2-Azabutane and 1-Thiapentane, respectively.

Now, look at the compounds below:-

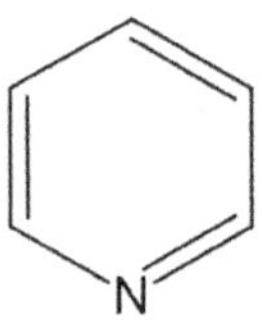

pyridine    7-methylidenecyclohepta-1,3,5-triene

In the two above structures, all of Huckel's rules are perfectly maintained, so they are aromatic compounds, but they don't have a benzene structure, so they are considered as Non-Benzenoids.

Let us check some more non-benzenoids:-

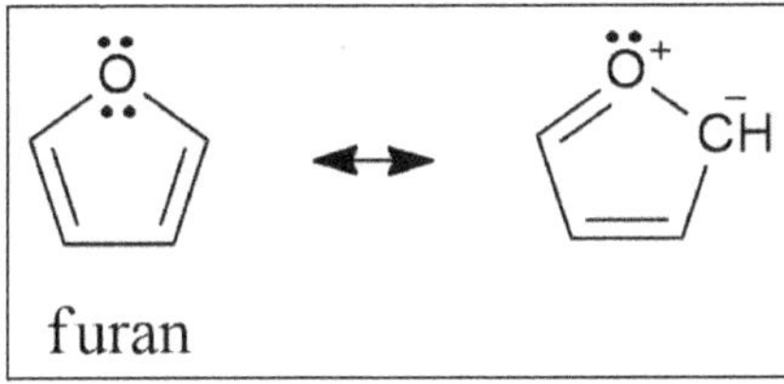

In furan structure, oxygen can only donate one electron pair and form a dative double bond.

So the number of electrons is 6, and it is a planar & cyclic compound.

This is also a non-benzenoid; you see the application of Huckel's rules of aromaticity in the final structures of resonance.

Non-Benzenoids are compounds having conjugated systems with a planar ring structure but do not have benzene rings in their structure. They may also contain heterocyclic compounds with Nitrogen, oxygen, etc.

What are the uses of these Non-Benzenoids?

Let us consider the compound pyridine; it is one of the most widely used compounds. It is used as an antiseptic in dental care, in antifreeze mixtures, as a reducing agent and disinfectant and in chemical industries as an important raw material.

It is used in the paint and dye industries and also used in coordination chemistry as a ligand.

Pyridine derivatives such as Pyridinium Chlorochromate (PCC) and Pyridinium Dichromate (PDC) act as oxidizing agents.

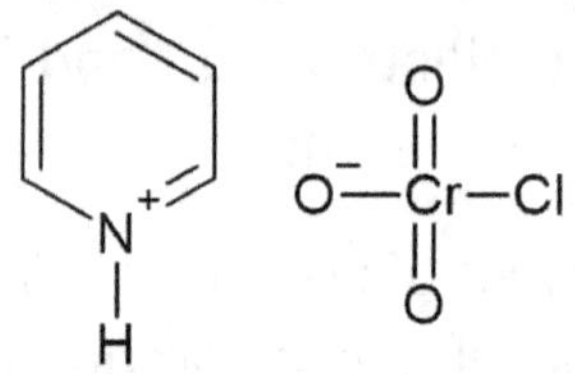

Pyridinum Chlorocromate

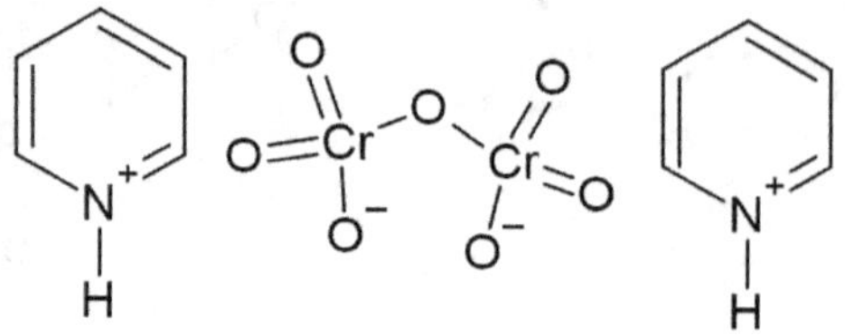

Pyridinium dichromate

These are the structures of the Pyridine derivatives; they are ionic compounds which are used in oxidizing organic compounds.

Chapter 12

The Geometry of Organic Compound

I have used the phrase "geometry of organic chemistry" previously; why is mathematics intruding on chemistry? I can understand the frustration you must be having. If in case you dislike me for that subject, I could have used the word "hate" instead of "dislike", but hatred is a strong word and should not be used frequently. Why am I even giving a lecture on language?

Coming back to the topic, According to many people, mathematics is just changing numbers based on the operation used. But some believe that math is nature or logic that defines nature, and there are many things that exist in this world for which there isn't any logic, for example, "why did newton write gravitational laws after he had seen the apple fall instead of making a pie?".

Well, in organic chemistry, we use geometry to predict the shape and alignment of the compounds, which helps in predicting their stability and existence. The shape of the compound depends on the number of bonds, types of bonds and lone pair on the atoms.

What are the different shapes possible in chemistry on the whole?

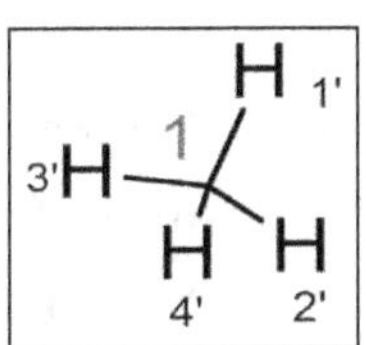

This is the actual structure of Methane; the structure is called tetrahedron structure. The structure is also known as a triangular pyramid; it consists of 4 corners and 4 triangular bases. Each corner is represented by a hydrogen atom, and carbon is present at the geometric centre of the structure.

This is how we represent the hydrogen on the Methane or any other compound. The thick line represents that particular hydrogen is on the front side or is facing the viewer or us. The dashed line represents that particular hydrogen is the rare hydrogen or the hydrogen, which is the backside.

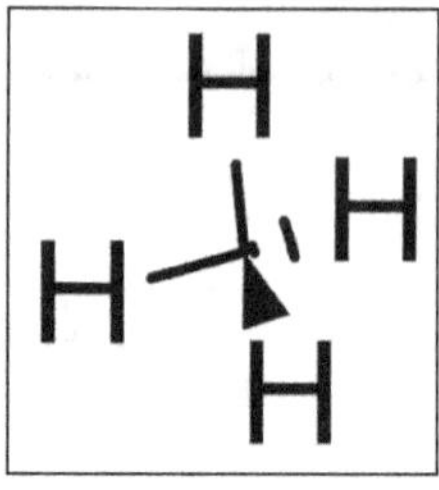

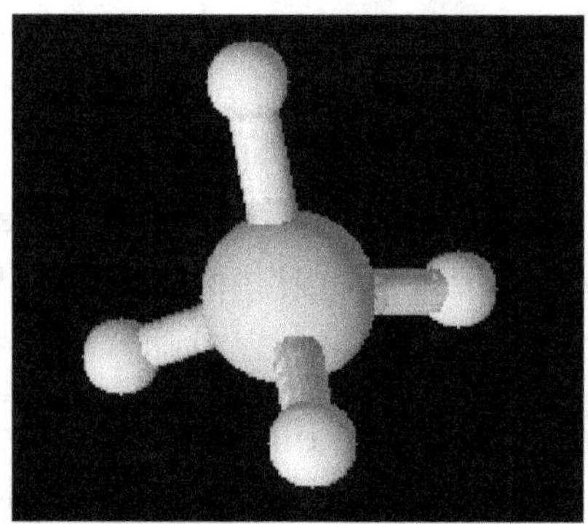

BALL and STICK Representation of Methane

This structural representation can be applied to any carbon compound. Now let's check for compounds like Ethane and Chloroethane.

In the above two images, both compounds have tetrahedron structures. But will have to observe it from a completely different point of view. In the ethane structure, first, consider carbon-1 to be at the centre in that situation, and consider hydrogen's (a',b',c') and carbon-2 to be the corner atoms.

And for the situation where carbon-2 is the centre corner, atoms are hydrogen (a,b,c) and carbon-1. The situation is similar with Chloroethane, except for the fact that one hydrogen atom will be replaced by a chlorine atom.

If we dive more into the topic of compound shapes, in organic chemistry, we only deal with carbon with tetravalency, and its structure is mostly tetrahedron or linear. But if you go through inorganic compounds, you can observe many shapes and sizes. (Well, sizes include bond length and bond angle, which will be further discussed). Time being, why don't we check a different kind of shape!!

This is the molecule of Phosphorus pentachloride, and its shape is trigonal bipyramidal.

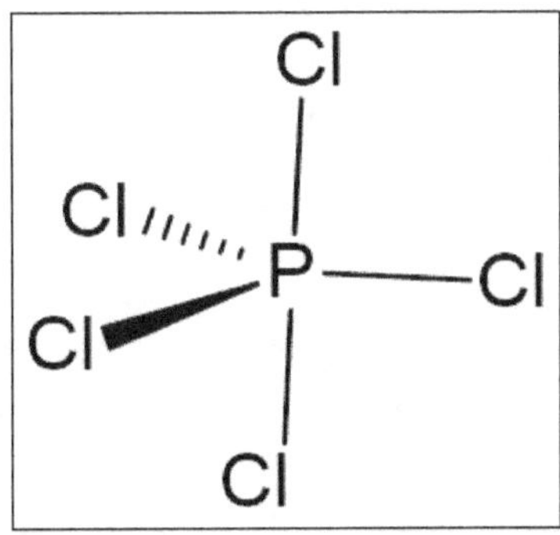

There are many other shapes in inorganic chemistry like pyramidal, planar, linear, bent, trigonal planar, octahedral and etc. But why don't we save them for another day?

Now, we will learn about hybrid orbitals and hybridization. First of all, what is an orbital? In your study of education, you might have come across orbitals in atoms; you must have read of s,p,d,f orbitals in fillings of electrons and the electron capacity of each orbital. Now, we will learn about the shapes of these orbitals and the formations of different bonds.

In physical chemistry, if you have gone through the topic of atomic structure, you will notice that number of electrons that can be in a single orbital will be increasing as follows s orbital can fit up to 2 electrons, p orbital can fit up to 6 electrons, d orbital can fit up to 10 electrons, and then f orbital can hold up to 14 until now that is the humankind achieve or observe on this plant.

Why don't we observe orbitals in 3D-plan and discuss them in detail?

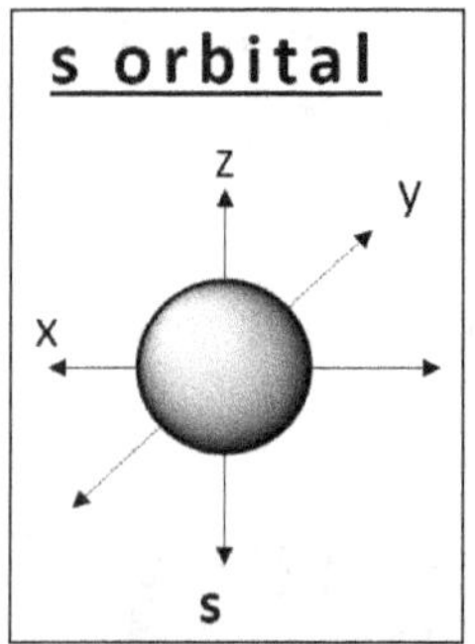

The shape of the s orbital is a spherical plane; the number of the electron in this orbital is 2. In all situations, the s orbital only forms a sigma bond; it is not involved in pi bonds. The sigma bond which is formed can be s-s overlapping, s-p overlapping, and head overlapping of p-p orbital will result in the formation of the sigma bond.

This is the shape of the p orbital; the p orbital has 3 sub orbitals, each on a different axis. Each suborbital can hold up to 2 electrons, and since

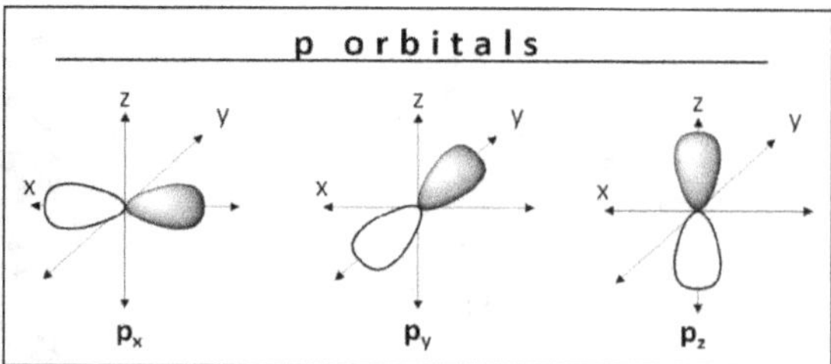

there are 3 sub orbitals, the total electrons are 6 in the p orbital. The lateral overlapping of p-p orbitals results in a pi bond.

As soon as the number of electrons and shapes of orbital differ from the previous one.

Now let us observe the formation of the sigma bond:-

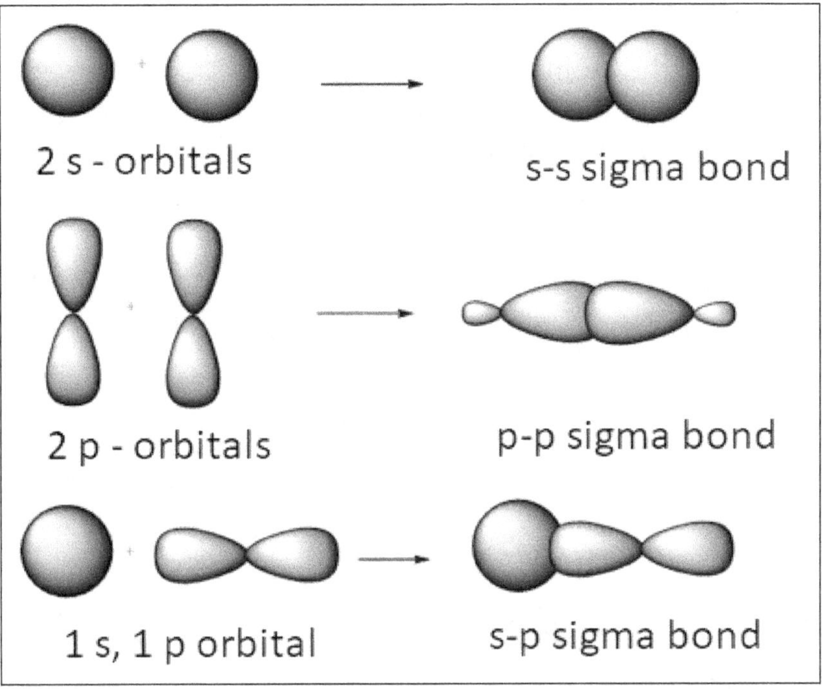

Let us observe the formation of sigma bond:-

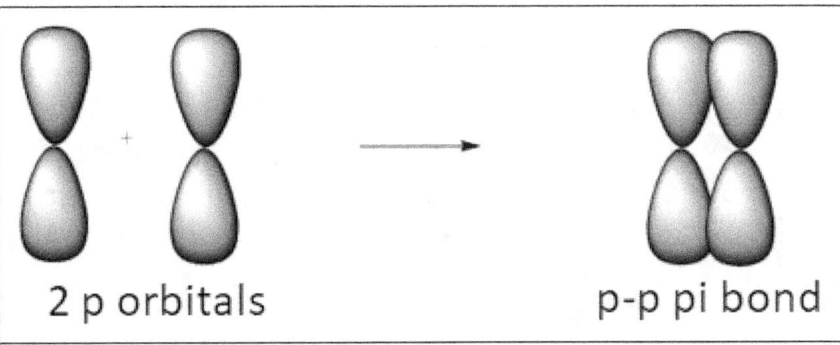

Since we have discussed the formation of different types of bonds, let us now check how the bond formation takes place in a carbon atom.

On the topic of atomic structure, you must have discussed the filling of electrons their energy levels; you must have seen the way we place electrons in the orbitals following Pauli's exclusion principle, Hund's rule and etc.

We know that the atomic number of carbon is 6 and the electronic configuration is $1s^2\ 2s^2\ 2p^2$. From this, we could understand that the first orbital or K shell is completed, but not the L shell, so we will discuss it in detail.

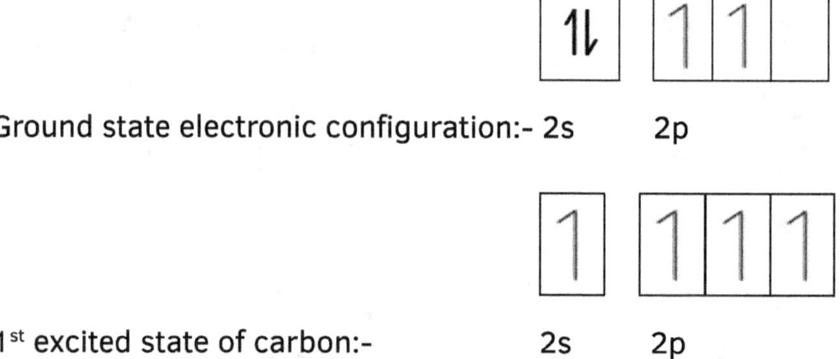

Ground state electronic configuration:- 2s   2p

1st excited state of carbon:-   2s   2p

In this excitation, the electron pair in s orbital has been split to form four unpaired electrons, which is the reason for carbon's tetravalency. Now let us see the formation of the bond using hybrid orbitals.

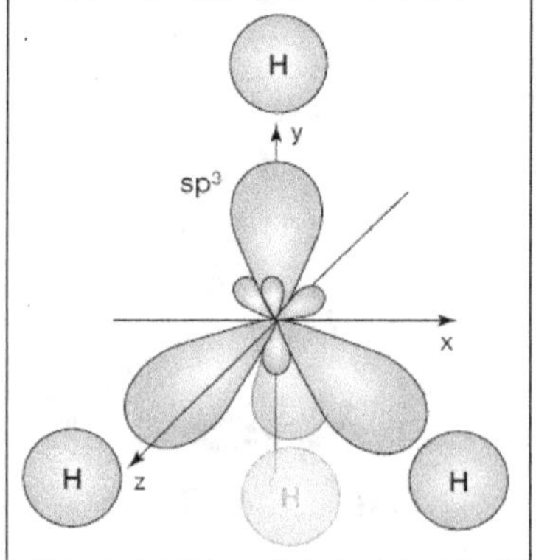

Hydrogen is a with one electron in 1s orbital; four hydrogen atoms form a bond with 1 s & 3 p orbitals forming sp³ that is one s orbital, and 3 p orbitals of

carbon are involved in bonding. The types of bonds in Methane are s-s, 3 s-p bonds.

In this compound, there are 3 sigma bonds that are 1 s, 2 p orbitals forming a bond with two hydrogens and the second carbon, whereas the third p orbital in both carbon atoms for lateral overlapping, which in turn creates and pi bond in the molecule.

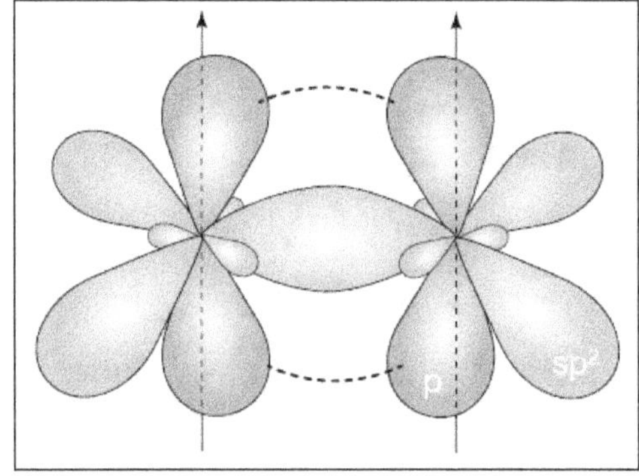

The same lateral overlapping, when done by two p orbitals from each carbon atom then it forms two pi bonds between them. The compound with such bonds is called alkyne.

Classification of Isomers

What are isomers?

The compounds with the same molecular formulae (condensed structure) but different physical properties or distinct arrangements of functional groups.

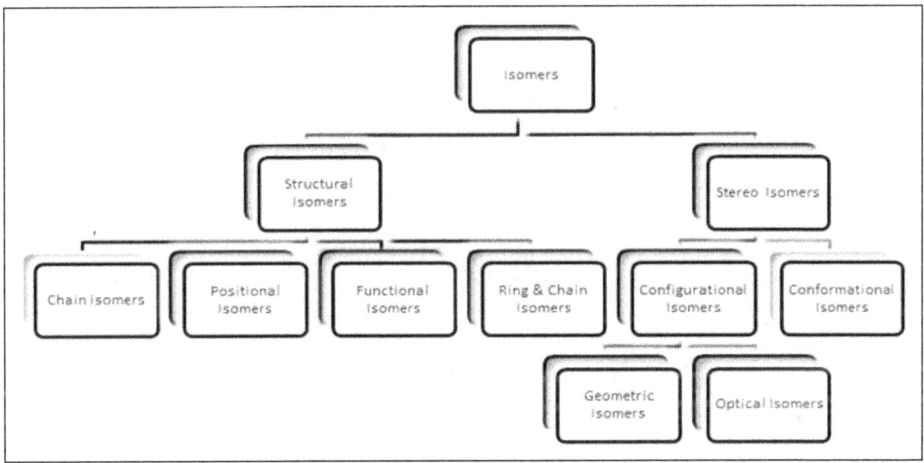

Let us discuss each of the structural isomers in detail:-

1. **Chain Isomers:-** The compound which has the same molecular formula but they have a different parent chain.

Ex:-

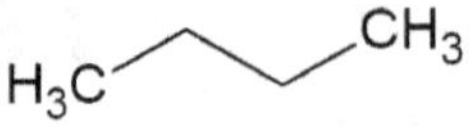

butane	2-methylpropane

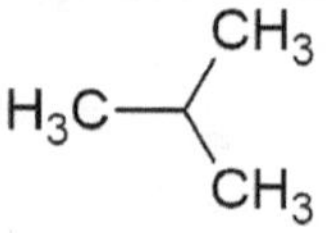

Butane and 2-methylpropane are chain isomers because they have the same condensed formula as C_4H_{10}, but the parent chain of both of them is different.

2. **Positional Isomers:-** The compounds with the same molecular formula and functional group but with different positions.

Ex:-

1-chlorobutane	2-chlorobutane

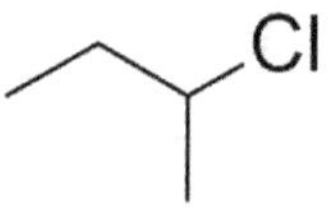

3. **Functional Isomers:-** The compounds with the same molecular formula but different functional groups.

Empirical Formula	Compound-1	Compound-2
$C_nH_{2n+2}O$	R—OH Alcohols	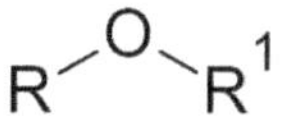 $R-O-R^1$ Ethers
$C_nH_{2n}O$	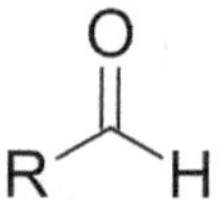 R—CHO Aldehydes	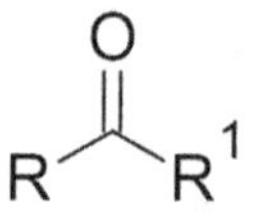 $R-CO-R^1$ Ketone

$C_nH_{2n}N$	$R{-}NH_2$ $R{-}\overset{NH}{R^1}$ $R{-}\underset{R^2}{N}{-}R^1$ 1°-amine 2°-amine 3°-amine	
$C_nH_{2n}O_2$	$R{-}\overset{O}{C}{-}OH$ Carboxylic Acid	$R{-}\overset{O}{C}{-}O{-}R^1$ Ester
$C_nH_{2n-1}N$	$R{-}{\equiv}N$ Cyanide	$R{-}\overset{+}{N}{\equiv}\overset{-}{C}$ Iso-Cyanide
$C_nH_{2n+1}NO$	$R{-}\overset{O}{C}{-}NH_2$ Amide	$R{-}\overset{OH}{=}N$ Oxime
$C_nH_{2n+1}NO_2$	$R{-}\overset{O}{\overset{+}{N}}{-}O^-$ Nitroalkane	$R{-}O{-}N{=}O$ Alkyl Nitrile

4. **Ring & Chain Isomers:-** The compounds with the same molecular formula, but the compound has a different arrangement or structure.

Ex:-

prop-1-ene   cyclopropane

This brings us to another important part of structural isomers, that is Metamers, the compounds with the same molecular but the number of carbons in the alkyl chain connected to the functional group is different.

This phenomenon can be observed in compounds like esters, amines, amides, ketones, esters and Thioethers.

Let us check Metamers for ether:-

Ethyl butyl ether:

Metamers of ethyl butyl ether	
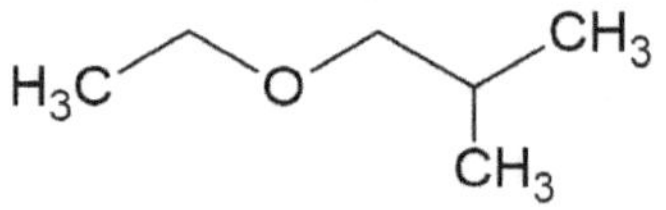	Ethyl 2-methylpropyl ether
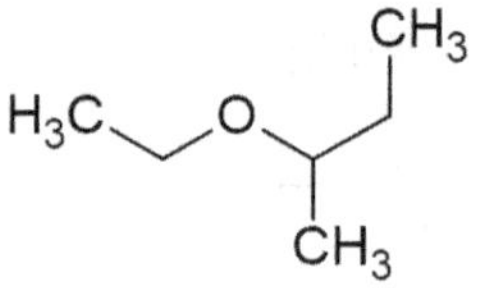	Ethyl 1-methylpropyl ether
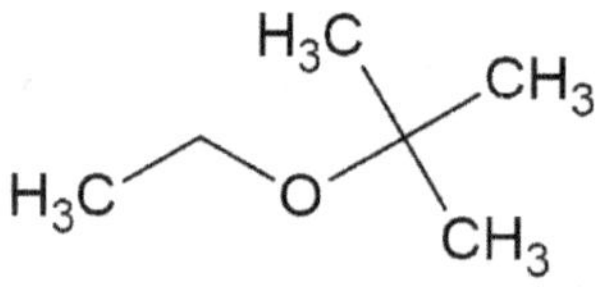	Ethyl tert-butyl ether
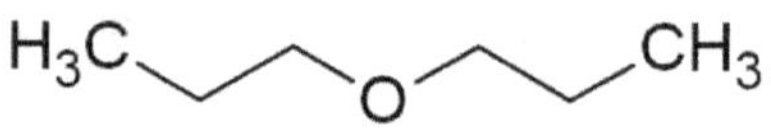	Di-n-propyl ether
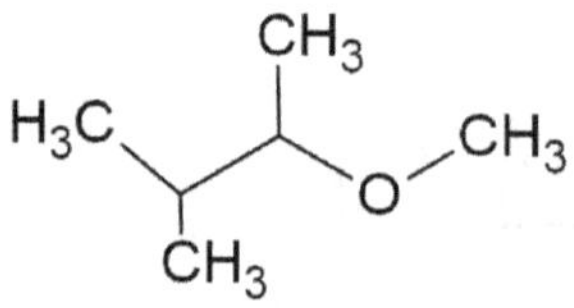	2,3-Dimethylbutyl methyl ether

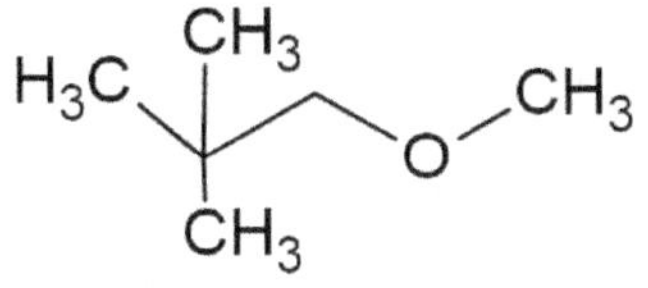	3,3-Dimethylbutyl methyl ether
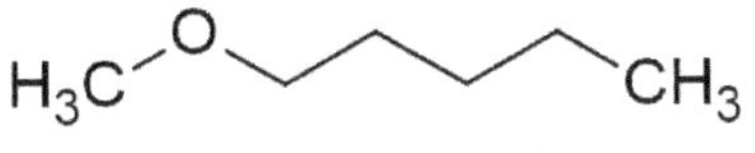	Methyl pentyl ether
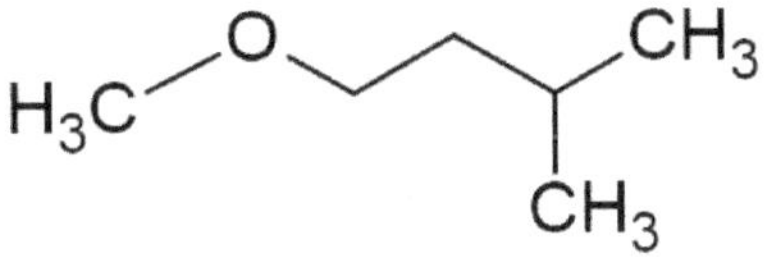	3-methylbutyl methyl ether
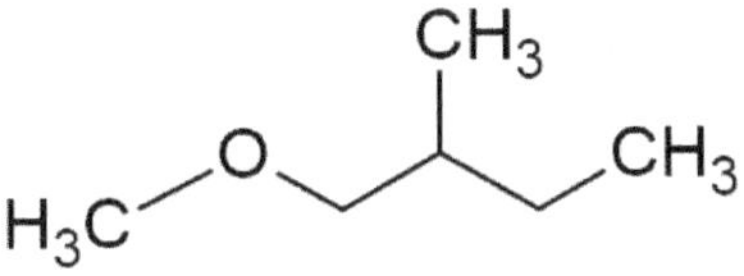	2-methylbutyl methyl ether
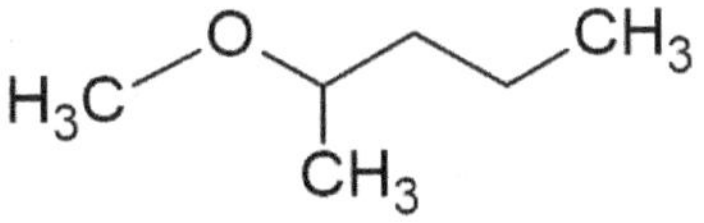	1-methylbutyl methyl ether

These are all the structural isomers for ethyl butyl ether; why don't you try for other compounds like pentan-1-amine, ethyl propyl ester?

So many isomers for a single compound with one functional group, how many are possible if we use more than one functional group? I will leave that to you.

Previously we have discussed ring and chain isomers. In such cases, how do we calculate the number of pi bonds and ring structures?

This question arises on the topic of saturation, how many carbons are completely saturated or have sp^3 hybridization or how many carbons have unsaturated that number carbons with pi bonds or if it is a cyclic compound?

This method is called "Degree of Unsaturation".

In analytical organic chemistry, the degree of unsaturation is used to calculate the total number of pi-bonds and rings in the compound. In general degree of unsaturation is a mathematical formula that will help find the number of multiple bonds and rings in a compound.

Why don't you check the condensed formula of cyclic butane and butene? You will observe that both of them have the same condensed formula, which is C_4H_8.

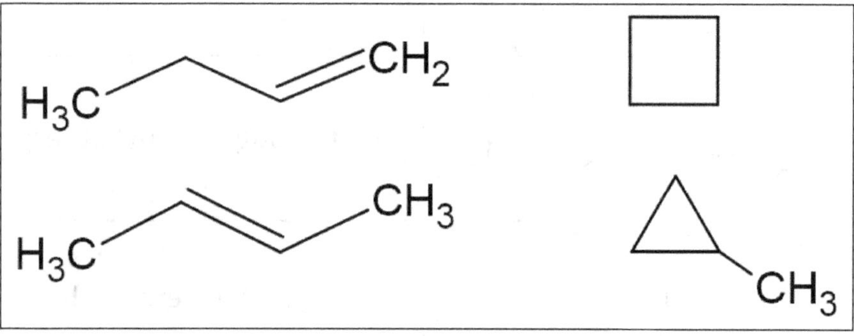

In all the above, the number of carbons is 4 and hydrogen's 8; we can say that all of these are ring and chain isomers. For this compound's degree of unsaturation is 1, we can also observe that in all compounds, there is either the pi bond or the ring structure, which means that one unsaturation value can either have one pi bond or the ring structure.

But how do we calculate the degree of unsaturation?

A formula was developed to calculate the degree of unsaturation. It is based on the valences of carbon, hydrogen, Nitrogen and halogens. The atoms like oxygen and other divalent atoms do not contribute to the degree of unsaturation, so they are not considered in the formula.

Degree of Unsaturation (DoU) $= \dfrac{(2C+2)-(H+X-N)}{2}$

C is the number of carbon atoms

H is the number of hydrogen atoms

X is the number of halogen atoms

N is the number of nitrogen atoms

Why don't you apply this formula to the previous compound that is C_4H_8?

The number of carbons is 4, the number of hydrogen is 8, and the number of halogens and Nitrogen is 0. The degree of unsaturation is as follows:-

$$DoU = \dfrac{(2(4)+2)-(8+0-0)}{2} = \dfrac{10-8}{2} = 1$$

Let us try for the compound $C_5H_9FClNO_2S$:-

Number of carbon atoms 5 Number of hydrogen atoms 8

Number of Halogen atoms 2 Number of Nitrogen atoms 1

Divalent atoms have no effect.

$$DoU = \dfrac{(2(5)+2)-(9+2-1)}{2} = \dfrac{12-10}{2} = 1$$ for both of these compounds,

we have obtained the degree of unsaturation is 1, but it does not mean that for all compounds, we will have the same degree of unsaturation.

This is a brief explanation of the degree of unsaturation; the derivation of this formula is nonessential in the current state of affairs.

Well, until now, we have final dig a little into the topic of organic chemistry.

I have a task for you; I want you to write your name using straight lines or without curves. Then assume those lines to bonds between carbons like in bond line structures and name the organic compound formed by each letter in your name.

If you want, you can add branches of halogens and other functional groups; see later A and Z given below for examples.

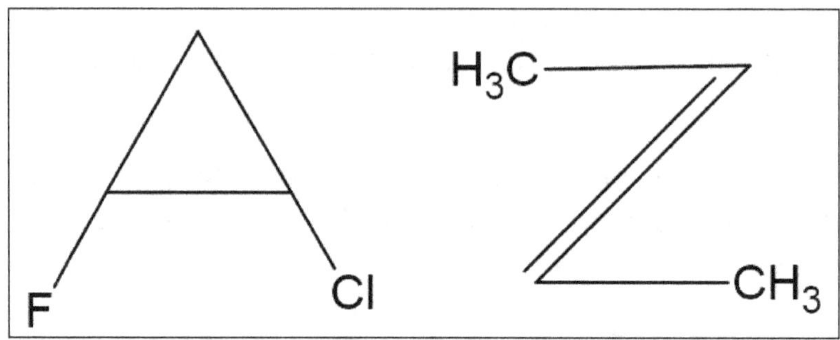

If you had fun naming your name, try naming your family member and then create a complex compound of your own and name it with the knowledge you have currently. Do not worry if you are wrong because mistakes and rejections make you a better person.

You might be worrying that ending is close, but we haven't discussed anything in the area of stereoisomers, nor I have said anything about conformational isomers?

You might have many questions in mind.

What are they?

How to learn or understand them?

Where are they used?

Are they important?

Well, I will answer the last question "they are important and an integral part of organic chemistry", but I am not going to discuss them now. With your permission I would like to take a leave, wait! Before leaving, I have another question.

Did you ever wonder why the C_2 compound does not exist like other compounds like O_2, F_2, etc.?

Let us discuss all of these some other day. Have a good day.............

-----------------------The end----------------------